ATOMS IN THE FAMILY

My Life with Enrico Fermi

by **LAURA FERMI**

The University of Chicago Press
Chicago & London

The University of Chicago Press, Chicago 60637
The University of Chicago Press, Ltd., London
© 1954 by The University of Chicago
All rights reserved. Paperback edition 1961.
Printed in the United States of America

01 00 99 98 97 96 95   20 19 18 17 16 15 14

ISBN 0-226-24367-2

♾ The paper used in this publication meets the minimum requirements of the American National Standard for Information Sciences—Permanence of Paper for Printed Library Materials, ANSI Z39.48-1984.

CONTENTS

CONTENTS

LIST OF ILLUSTRATIONS

vii

On the campus of the University of Chicago there is an old, dilapidated structure, an imitation of a medieval castle with turrets and battlements. It is only a façade, concealing the west stands of a football stadium that is no longer used. The stucco walls are coated with thick soot; huge chimney pipes emerge from windows and reach above the battlements.

Sightseeing busses stop in front of this structure. Guides direct tourists to a plaque on the outer wall:

ON DECEMBER 2, 1942
MAN ACHIEVED HERE
THE FIRST SELF-SUSTAINING CHAIN REACTION
AND THEREBY INITIATED THE
CONTROLLED RELEASE OF NUCLEAR ENERGY

This is the birth certificate of the atomic era.

The first atomic pile ever made was erected by a small group of scientists in a squash court under the football stands. They worked in great secrecy, at their fastest pace, pressed by the urgency of their aim. The second World War was being fought. The men in the squash court knew that their research might make possible the development of atomic weapons.

The scientists operated their pile for the first time on December 2, 1942. They were the first men to see matter yield its inner energy, steadily, at their will. My husband was their leader.

This book is the story of his life and mine, of the work that made possible the squash-court experiment, of events that came before and after it. The story starts in Italy, where I was born and where I lived the first thirty-one years of my life.

ix

Publisher's Note:

The west stands of the football stadium to which Mrs. Fermi referred were razed in 1957. Henry Moore's sculpture *Nuclear Energy*, which was dedicated on the twenty-fifth anniversary of the birth of the atomic age, now marks the location of the first self-sustaining chain reaction.

PART I

ITALY

FIRST ENCOUNTERS

On a spring Sunday in 1924 a group of friends asked me to join them for a walk, and we met at a certain streetcar stop on a certain street of Rome. Along with my friends came a short-legged young man in a black suit and a black felt hat, with rounded shoulders and neck craned forward. In Italy a black suit means mourning for a close relative, and I learned later that his mother had recently died. His hair was also black and thick, his complexion dark. In introducing him, my friends tried to impress me:

"He is a promising physicist, already teaching at the university, although he is only twenty-two."

To me this explained the young man's posture and queer appearance; but twenty-two seemed pretty old and time enough for achievement: I was sixteen.

He shook hands and gave me a friendly grin. You could call it nothing but a grin, for his lips were exceedingly thin and fleshless, and among his upper teeth a baby tooth lingered on, conspicuous in its incongruity. But his eyes were cheerful and amused; very close together, they left room only for a narrow nose, and they were gray-blue, despite his dark complexion.

"We want to be in the open, not among houses," my friends said.

The countryside around Rome is beautiful and easy to reach. One may go west on the electric train to the deep-blue Tyrrhenian sea and its burning sands; or south on the outmoded Vicinali railroad to the numerous towns perched on the hills around Rome. One may simply ride a streetcar or a bus to the end of its route and soon find one's self in a little vale where a brook murmurs in the shade of oaks and beeches; or on an ancient Roman road flanked by sun-

3

baked ruins and umbrella pines; or on the top of a rocky hill, in the cool peace of an old monastery, half-hidden among dark cypresses.

We rode a streetcar on that Sunday afternoon, and a short walk from its terminal took us to a large green meadow near the confluence of the Aniene River with the Tiber, a zone that is now entirely built up with crowded apartment houses. Naturally, as if it were his role, the young physicist took the lead, marching ahead of the group, his neck craned forward, his head apparently more anxious to reach places than his feet.

"We'll play soccer," he stated.

I had never played soccer in my life and I was no tomboy, but he had spoken. There was no opening left for argument, and no opportunity for complaint.

The game must have been planned before, for my friends produced a deflated soccer ball, which was soon filled, with everybody's mouth helping in turn to blow it up. We divided into two teams, and I was in that led by the young man with the black suit.

"What am I to do?" I asked discouragedly.

"You shall be goalkeeper; it is the easiest task. Just try to catch the ball when it goes through the goal. If you miss it, don't worry; we shall win the game for you." The young man's attitude toward me was protective.

There was an easy self-reliance in him, spontaneous and without conceit. Luck, however, was against him: at the height of the game the sole of one of his shoes came loose and dangled from the heel. It hampered his running, made him stumble and fall on the grass. The ball zoomed above his fallen body and sped toward the goal. It was up to me to save the day: while I was observing our leader's predicament with more amusement than pity, the ball hit me in the chest. Stunned I wavered, almost fell, recovered my balance. The ball bounced back into the field and victory was ours.

Our leader pulled out of his pocket a large handkerchief, wiped off the perspiration that streamed down profusely from the roots of his hair over his face, then sat down and tied his loose sole with a piece of string.

That was the first afternoon I spent with Enrico Fermi and the only instance in which I did better than he.

After that first afternoon I did not see Fermi for over two years, until we met again in the summer of 1926. Our second encounter was due to Mussolini.

My family had planned to spend the summer in Chamonix, a resort in the French Alps, on the slopes of Mont Blanc. Consideration of the advantageous rate of exchange had induced my parents to take the momentous decision of going to a foreign country. Our passports had been secured without much difficulty, because my father, an officer in the Italian navy, had been able to pull a few strings. Hotel reservations were made, and we were ready to pack. Then my father came home with the news that no foreign currency was available on the Italian market and that very recent restrictions would prevent us from taking lire to France. *Il Duce* was quietly preparing his financial policy, the "battle for the lira," which he officially announced a month later, in August, 1926, when he gave his well-known speech at Pesaro. This sudden closing of the frontiers to the nonessential flow of money was the first step toward the strict monetary controls that kept the lira at an artificially high value throughout Fascist times, creating an economy which required more and more regulations and increasingly tighter controls.

When my father brought home word of these first restrictions, he was unable to explain them, and we could not understand Mussolini's reasons for not letting us spend a summer in Chamonix. But my father allowed no criticism: in the navy he had been trained to consider authority necessary for the smooth functioning of human society and to respond to authority with discipline and obedience. The period of unrest after the first World War had caused great concern to him: public demonstrations against the government, strikes, seizure of factories, threats of communism, were against the principles under which he had been raised and by which he lived. Mussolini was to him the strong leader whom Italy needed if order and moral rectitude were to be restored. My father never doubted that, once the *Duce* had achieved this aim, he would gradually revert to a more democratic form of government.

When we children started to grumble over our upset plans, my father put a stop to our recriminations:

"*Il Duce* knows what he is doing. It is not for us to judge his actions. There are numberless resorts in Italy as good as Chamonix or even better. We'll go somewhere else, that's all."

In Italian families of that time decisions were made by parents alone. A girl who had just become nineteen had little chance to voice her opinions; so it was in a shy and subdued tone that I advanced a suggestion:

"Why don't we go to Val Gardena? The Castelnuovos will be there. . . ."

Professor Guido Castelnuovo was a mathematician with many children, some of them my friends, the same who had been in the group playing soccer two years before. Wherever the Castelnuovos went, a number of other families were sure to follow.

My parents looked at each other and smiled. They must have seen in a flash that most picturesque valley of the Dolomites, which climbs up to the massive rocks of the Sella group and widens here and there into a sunny basin cradling a little village with its red roofs and aerial church steeple.

"Do you still remember the delightful summer we spent in Selva?" my father asked my mother in that faintly nostalgic mood that accompanies pleasant recollections. I knew at once I had sold my idea.

"We could go there again," my mother said, "or, even better, we could go to Santa Cristina. It's in a prettier spot and offers a better choice of hotels."

By the middle of July we arrived in Santa Cristina. The Castelnuovos were staying in the village right below, and I went to see them. Gina, the closest to me in age, was full of expectations for the summer.

"We are going to have lots of fun. So many people will come here. Even Fermi has written my mother asking her to find a room for him."

"Fermi?" I queried, "Fermi? . . . The name sounds familiar. . . ."

"You must know him, I am sure. He is a brilliant young physicist; the hope of Italian physics, as my father says."

"Oh yes! I remember him now: he is the queer guy who made me play soccer. I had entirely forgotten his existence. Where has he been hiding himself all this time?"

"He was in Florence, teaching at the university there. But in the fall he will come to Rome."

"To Rome? What's he going to teach?" I was then a student in general science at the University of Rome and was required to take, among others, courses in physics and mathematics.

"The faculty of science has established a new chair *ad hoc* for Fermi: theoretical physics. I think that Corbino, the director of the physics laboratory, had a great deal to do with calling him to Rome. Corbino holds him in very high esteem and says that of men like Fermi only one or two are born each century."

"This is certainly an exaggeration," I interrupted. The young physicist had made no impression on me. Among my school friends there were boys who seemed more brilliant and promising to me. "Anyway, theoretical physics is not a subject that I am going to take, so Fermi will not be my teacher. What's he like as a friend?"

"Grand! My father and the other mathematicians would like to talk shop with him, but he runs away as soon as he can and comes to us. He enjoys games and hikes and is very good at planning excursions. Besides, my mother trusts him and lets me go anywhere if he is along."

Soon I was to learn by personal experience that Fermi was fond of physical exercise.

"We must get into shape fast. Tomorrow we shall take a short walk, and the day after a longer one. Then we shall start climbing mountains," he stated as he made his appearance in Val Gardena. In knickerbockers and short Tyrolean jacket, he looked more natural and less queer than the first time I had seen him.

"Where shall we go?" Cornelia asked. She was a sturdy woman, the sister-in-law of a mathematician friend of the Castelnuovos, Professor Levi-Civita. She was bubbling with energy and anxious to be on the go.

Fermi was already bending over a map.

"We can hike in Valle Lunga ["Long Valley"] up to its top."

"How long a hike is it?" Gina asked.

Fermi put his thick thumb on the map and moved it up several times to cover the distance between the bottom and top of the valley. Fermi's thumb was his always ready yardstick. By placing it near his left eye and closing the right, he would measure the distance of a range of mountains, the height of a tree, even the speed at which a bird was flying. Now he mumbled some figures to himself, then answered Gina's question.

"Not very long. About six miles each way."

"Six miles! Isn't it too much for the young children who want to come along?" Cornelia asked. Our group, the Castelnuovos with their cousins and various friends, included youngsters of all ages: the predominance of family friendships over those made in school helped to avoid the separation into age groups prevalent in the United States.

Fermi turned to Cornelia with mock sternness:

"Our new generations must grow strong and enduring, not sissy. Children can walk this much and more. Let's not encourage them to be lazy!"

No more objections were raised. It was always thus: Fermi would propose, and the others would follow, relinquishing their wills to him.

He was not twenty-five years old, but he had already the earnest look of the scholar and the assurance of the man used to exerting an ascendancy through teaching and advising young people. At first sight he gained my mother's confidence, and I, too, like Gina, was allowed to go on excursions planned by him. My parents never questioned his judgment, the length of the hike, or the difficulty of the climb. They only insisted that my young brother or one of my two sisters go along, because decorum was the foremost of their concerns.

We would leave at dawn, carrying our knapsacks on our shoulders. Fermi's was always the bulkiest and heaviest of all: he stuffed it with the lunch and sweaters of any child who went with us and, during a steep climb, also with the pack of any girl who seemed to be tired. He took pride in the size of his sack, which bent his broad shoulders at the slant of the hill; but, forgetful of the large bulge on his back, he would wag it right and left, hitting whomever he was trying to pass at the moment. Fermi was passing people frequently: whenever the trail became steep, he deemed it his duty to run to the head of the line of hikers and assume the role of mountain guide:

"Copy my steps if you want to save yourselves some trouble."

Many would lag behind.

Every half-hour, or so, Fermi would stop, sit on a rock, and announce:

"Three minutes' rest."

By the time all the others had reached him, Fermi would be on his feet again.

"We are all rested. Let's go."

Nobody dared protest. But once Cornelia, who, being a little older than the rest of us, was less restrained, turned to him and asked:

"Are you *never* out of breath? Does your heart never jump in your throat?"

"No," Fermi answered with his most modest grin, "my heart must have been custom-made; it is so much more resistant than anybody else's."

When approaching the end of our climb, Fermi would pass any of us who happened to be ahead of him, for he could not possibly allow anybody to beat him and reach the top before he did. On his short legs he jumped from rock to rock, wagging his knapsack against us, until he had everybody behind.

Filled with the joy of the conquered height, relaxed now that the climb was over, we always spent some time on the top. From that high the view of the Dolomites always appeared fantastic, with improbable shapes of towers and pinnacles, the distant glaciers, the close-by perennial snows. The elation experienced on top of a mountain is unique and always new. There is a moment of reverent silence, a subconscious identification of the individual with nature and divinity, an instant of pantheistic worship. Then comes the merriment, the excited chatter, the exchange of impressions, the singing of mountain songs. And the regret at having to climb down.

At lunch time we used to look for a soft meadow, the shade of a tree, and a brook from which to draw the crystal-clear water. After eating, we would lie on the grass and perhaps doze. Fermi would suddenly call to us:

"See that bird there?"

"Where?"

"On the top branch of that big tree high on the hill. Perhaps you mistake it for a leaf at this distance."

9

Nobody could see the bird.

"My eyes must have been custom-made; they see so much farther than anyone else's." Fermi was apologetic, as if begging forgiveness of his friends who had to be satisfied with mass-produced eyes.

All parts of his body were custom-made and better than other persons': his legs were less easily tired, his muscles more resilient, his lungs more capacious, his nervous system steadier, and his reactions more accurate and prompt.

"What about your brains?" Gina asked him once in a teasing mood. "Were they also custom-made?"

On this point, however, Fermi had nothing to say. He had no interest in his intellectual powers. They were a gift from nature over which he had less control and in which he took less pride than in his physical abilities. But to intelligence in general he gave much thought. Although he often claimed that intelligence is an indefinable entity, made up of many factors that are not easy to evaluate, still one of his favorite pastimes during that summer was to classify people according to their intelligence. Fermi's love for classification was inborn, and I have heard him "arrange people" according to their height, looks, wealth, or even sex appeal. But that summer he limited his classifications to the intellectual powers.

"People can be grouped into four classes," Fermi said: "Class one is made of persons with lower-than-average intelligence; in class two are all average persons, who, of course, appear stupid to us because we are a selected group and used to high standards. In class three there are the intelligent persons and in class four only those with exceptional intellectual faculties."

This was too good a chance of teasing Fermi to let it go.

"You mean to say," I commented, assuming the most serious expression I could manage, "that in class four there is one person only, Enrico Fermi."

"You are being mean to me, Miss Capon. You know very well that I place many people in class four," Fermi retorted with apparent resentment; then he added on second thought: "I couldn't place myself in class three. It wouldn't be fair."

I did not give up but went on nagging at him. By way of protest he ended by saying:

"Class four is not so exclusive as you make it. You also belong in it."

He might have been sincere at the time, but later he must have demoted me to class three. Be that as it may, I have always liked to have the last word in any argument, and so I said with finality:

"If I am in class four, then there must be a class five in which you and you alone belong." To everyone except to Fermi, my definition became a dogma.

(2)

THE TIMES BEFORE WE MET

Next fall Fermi settled in Rome for good. He went to live with his father Alberto and his sister Maria in a small house in Città Giardino.

Città Giardino, or "Garden City," would be termed here a government-subsidized housing project for medium-income government employees. It was built between 1920 and 1925 on an area a few miles northeast of Rome. At that time it consisted exclusively of small, one-family homes surrounded by gardens. Tenants paid a moderate rent and became owners in twenty-five years.

The northern edge of Città Giardino was reserved for railroad employees. Fermi's father, who was one of them, obtained a house in that section and moved into it with his daughter in the fall of 1925. Both Fermi's parents had looked forward to their new home, but Mrs. Fermi never saw it finished, for she had died in the spring of 1924, and Mr. Fermi enjoyed it but a short time, until 1927, when he, too, passed away.

I was to set foot in that house only after I became engaged to Enrico Fermi early in 1928, but before that I had seen it from the outside. Pushed by a nonadmitted interest in the Fermis' whereabouts, I had once explored Città Giardino and walked down Via Monginevra. The Fermis' residence was at No. 12, at the foot of a hill and just above the depression in which the Aniene River flows before reaching the confluence with the Tiber. A low brick wall, topped by a spiked iron rail, ran along the street. Some newly planted ivy struggled hard to climb the iron rail. The house was a few feet back of the wall, and the garden extended farther back on a steep slope. The house would have been plain but for a squat tower of some sort over its flat roof.

Inside it was neither large nor luxurious, but comfortable enough. It had hot water in the bathroom and represented a great improvement over the apartment in which the Fermis had lived since 1908, in a large building at Via Principe Umberto 133, near the railroad station.

All houses in the section near the station were hurriedly built soon after Rome was united to the Kingdom of Italy and made its capital in 1870. Then the sudden affluence of employees who had moved south from Piedmont with the government had created an acute housing shortage that gave a free hand to speculators. For all its pretense of elegance—two statues were in the hall at the feet of two ample stairways—the building at Via Principe Umberto 133 had provided none of the niceties of modern living. There was no heating of any kind, and the three Fermi children, Maria, Giulio, and Enrico, often had chilblains in the winter. Enrico still likes to tell "the soft younger generations" how he used to study sitting on his hands to keep them warm and how he would turn the pages of his book with the tip of his tongue, rather than pull his hands out of their snug warming place.

The apartment in Via Principe Umberto also lacked a proper bathroom. It had a toilet only, and the family used two zinc tubs for their morning ablutions. The smaller one was for the children; the bigger tub, mounted on casters, was wheeled daily into the parents' room. Both tubs were filled with cold water at night and by morning had reached room temperature, which in winter may have remained below fifty degrees.

With never-questioned discipline the three children plunged into the cold water every morning, knowing that peasant stock like theirs could not be sissy.

The Fermis came from the rich countryside around Piacenza in the valley of the Po River, the most fertile land in all Italy. Enrico's grandfather Stefano was the first Fermi to abandon actual tilling of the soil, thus initiating the family's social rise. As a youth, Stefano entered the service of the Duke of Parma—Italy was still divided into small states, one of which was the Duchy of Parma—and obtained the position of county secretary. The brass buttons of his uniform, bearing the name and emblem of the Duke in relief, are still kept among family heirlooms.

Stocky and short-legged like all male Fermis after him, Stefano was of an astoundingly robust constitution and of a steely determination, hardly revealed by his mild blue eyes. His material positivism drove him to strive with unrelenting will power for a solid financial position without ambitions of grandeur. It was almost unavoidable that these traits should be associated with personal coolness, and he showed no undue tenderness for his numerous children, who were left to shift for themselves at an early age.

Enrico remembers him only vaguely, an old man bent in two by arthritis and made kindlier by age. He had become placid and benevolent and seemed to have one regret only, that his grandchildren would not drink wine with the relish of his own generation.

At his death in 1905, for all his thrift, Stefano had only a small inheritance to leave: a house and a small piece of land near the town of Caorso, where he had settled. The earthly goods left by him did not amount to much, but along with them went the mold in which he was shaped, and that brought great returns to the Fermi family.

Alberto, Stefano's second son and Enrico's father, was compelled to leave school sooner than his profound and eager mind deserved. But his father bade him look after himself and find his own means of livelihood. With no formal education, he entered the railroad administration.

Railroads, which had been slow to develop in Italy, were then in a phase of rapid expansion and provided excellent opportunities to capable men. Alberto brought into his work the qualities inherited from his father, steadiness, will power, determination to reach a modest prosperity. Soon he gained the respect and recognition of his colleagues and started to climb the ladder that led him to the position of division head, which he was holding at the time of his retirement and which was usually reserved for persons with university degrees.

His work caused him to wander all over Italy for several years and then to settle in Rome. In that city at forty-one he married Ida De Gattis, a woman fourteen years his junior, a teacher in the elementary schools. By her he had three children: Maria, born in 1899, Giulio in 1900, and Enrico on September 29, 1901. The three children came so close together that Mrs. Fermi could not care for the second and third babies and sent them to nurses in the country. Because of delicate health, Enrico was not returned to his family in Rome until he was two and a half years old, a fact that was jokingly recalled to him whenever he appeared to be "dumber" than the others.

Maria, though very young herself at the time, still vividly remembers her little brother's homecoming. He was small, dark, and frail-looking. The three children stared at each other for a while, then, perhaps missing the rough effusiveness of his nurse, little Enrico started to fuss and cry. His mother talked to him in a firm voice and asked him to stop at once; in this home naughty boys were not tolerated. Immediately the child complied, dried his tears, and fussed no longer. Then, as in later childhood, he assumed the attitude that there is no point in fighting authority. If *they* wanted him to behave that way, all right, he would; it was easier to go along with *them* than against. A losing cause is worth no effort.

Soon the child became not only adjusted to his family but also strongly attached to it. It was Mrs. Fermi's thorough and intelligent devotion to her husband and children that kept them close together. Her devotion was mixed with an overstressed sense of duty and an inflexible integrity, which the children inherited, although they occasionally resented it. Into her affection she brought a certain rigidity that made her expect from others as much as she would give. Her children were to work hard to maintain the high moral and intellectual standards that she had set for them and exacted of them.

In the winter of 1915 a dramatic occurrence struck the Fermi family and upset their spiritual balance. Giulio developed a throat abscess. It grew to the point that breathing became difficult, and the doctor advised surgery. It was to be only a minor operation, and the boy would be allowed to return home right after it. On the appointed morning Mrs. Fermi and Maria accompanied him to the hospital and set themselves to wait quietly in the hall. Suddenly there was a great commotion. Nurses rushed into the hall, saying aimlessly, "Don't worry; you should not worry." Their tone was strained. The surgeon came. He asked the women to keep calm. He could not explain, he could not quite understand himself what had happened. The boy had died before the anesthesia was completed. The blow could not have been heavier, nor the family less prepared to receive it.

In appearance the mother was the most deeply affected. Mrs. Fermi had shown a marked preference for Giulio over the other two children. All three were bright, and their parents did not evaluate their intellects beyond their school marks. From early childhood Giulio and Enrico, only a year apart in age, had grown so close to each other, had shared their games and filled their leisure hours together so incessantly, that it was hard to say what each of them brought into the partnership. The pair turned into wonder boys when just out of childhood. They built electric motors of their own design and made them work, they drew plans of airplane engines—all children were fascinated by the new invention—that experts said could not possibly be the work of children. They were indistinguishable in achievements.

But Enrico lacked some of the more likable traits of children. He was small for his age and unattractive. He was untidy, and his mother, when out with him, often made him stop to wash his face at a street fountain. His hair was never combed. He was uncommonly shy of grownups—when I met him and he was already twenty-two, young people considered him talkative, older ones taciturn—he went into tantrums easily and lacked imagination, or so it seemed. In school he did poorly in writing. Those qualities that were to become assets in his scientific prose—the going straight to the point with no flourishes, the simplicity of style, the avoidance of any word not strictly essential—were held to be symptoms of mental aridity in the child.

Once, when in second grade, he was to write what could be made with iron. Because on his way to school he used to pass by a store with the sign "Factory of iron beds," the child wrote only: "With iron one makes some beds." The sentence was clear and precise. By saying "some" the boy had implied his awareness that not all beds are made of iron. The second-grade teacher, however, was not pleased, nor was Mrs. Fermi, who came to doubt the depth of his intelligence.

Giulio, more affectionate and cheerful, more outgoing and responsive, had been his mother's favorite.

She never recovered from the blow of his death. She had been a lively, witty woman; now she was given to long spells of crying, and she let herself go into moods that were upsetting to her family, as she was too intelligent not to realize.

Mrs. Fermi found an outlet for her grief in her tears. But Enrico's mute sorrow may have been deeper than hers. His brother had been his steady companion and sole friend. There had been no need for others because the two completed each other to form a unit, as two atoms unite to form a molecule. They had no free valency to hook onto others. Now lonely, but shrinking from any show of feelings, he kept his grief to himself. A week after his brother's death he walked alone by the hospital where the fatal accident had occurred. He wanted to prove to himself that he was capable of overcoming the emotion that the sight of the hospital would arouse in him.

There was one thing that Enrico could do alone to fill the melancholy hours: study. And study he did, following an avid interest in science. He never gave up his outdoor activities, for he was thirteen and needed exercise. With the other boys in his class he went on playing ball and "French war," a game as popular in Italy as "Cops and robbers" is in the United States. He played detachedly for the sake of the game, for the boys were acquaintances rather than friends. And at home he gave himself to study for his own pleasure, not as a school requirement, needing little work to keep at the top of his class.

First he learned mathematics, then physics.

A major difficulty was the procurement of books. He had little spending money, and his father had no stocked bookcases, although he had acquired a vast culture by himself. Enrico Fermi became an assiduous frequenter of Campo dei Fiori, the well-known outdoor market held each Wednesday, where collectors used to unearth old books and prints, art objects, and all sorts of antiques. In Campo dei Fiori experts in the subtle art of bargaining could obtain anything at a reduced price, from fresh fish to flowers, from secondhand clothes to art treasures. Campo dei Fiori was a glorified Maxwell Street of Chicago, but, unlike Maxwell Street, it spread out on ancient streets around the monument to the burned-at-the-stake philosopher Giordano Bruno, with the background of the magnificent Palazzo Farnese built by Sangallo and Michelangelo.

It was not long before Fermi had a companion on his expeditions, another boy, Enrico Persico, who was to be my mathematics teacher at the university a full decade later.

One year older than Enrico Fermi, Persico had been in school with Giulio, whose prompt intelligence he had admired. He had not tried to become his friend, feeling that Giulio was "saturated" by his brother. In an accidental encounter after Giulio's death, Enrico Fermi and Enrico Persico realized that they had more than a first name in common. Their taste, their scientific aptitude, their proclivity to speculation, were similar. But in temperament they differed.

It has been said that the main characteristic of the Italian population is the variety of their noses. Since noses reflect temperaments, it seems safe to state that in Italians the variety of temperaments is greater than in other populations.

Fermi's nose, straight, thin, and pointed, indicated that he is forthright and self-assured but not arrogant; avid of knowledge but not unduly curious; that others will seldom succeed in changing his opinions but that he will not force his own on others.

Persico's nose told a different story. It started boldly forward, but with a sharp bend turned down abruptly to look at his feet. One recognized on sight that he was made for success but was hampered by numberless complexes; that he was retiring and cautious; that he would hide his positive traits under his shyness, as a violet will hide its color and perfume under green foliage.

Because they had similarities and contrasts, a friendship grew between the boy Persico and the boy Fermi that was to hold fast through the years. Meanwhile, they needed books to learn mathematics and physics, and Wednesday after Wednesday they patiently searched Campo dei Fiori. They acquired a few books and took turns in reading them.

On coming home after a purchase, Fermi showed it to his sister, whose inclinations were literary, philosophical, and religious, but definitely not scientific. Vainly he tried to make Maria share his own enthusiasm! Once he brought home from Campo dei Fiori a two-volume treatise on mathematical physics, and he told his sister that he was going to start reading it right away. During the next few days Maria was often interrupted in her studies by her brother's increasingly excited remarks about the book:

"You have no idea how interesting it is. I am learning the propagation of all sorts of waves!"

"It is wonderful! It explains the motion of the planets!"

His enthusiasm reached a peak when he arrived at the chapter on the recurrence of ocean tides. Finally, he got to the end of the book, and then once more he went to his sister:

"Do you know," he said, "it is written in Latin. I hadn't noticed."

The book was by a Jesuit physicist, a certain Father Andrea Carafa, S.J., and had been published in 1840. Both Enrico and Persico continued to hold the opinion that it was a good book.

As their knowledge of physics grew in vastness and profundity, the two friends set themselves to apply it to experimental problems. With the rudimentary equipment that they could procure, they made some accurate enough measurements, that of the magnetic field of the earth, for instance. They also tried to explain a certain number of natural phenomena, and for a long time they were puzzled by what seemed to them the deepest mystery of nature: like most other boys, they used to play with tops, which were popular because not expensive. Unlike most boys, however, they had tried to explain the top's strange behavior. That it should spin faster the more violently they pulled away the string wound around it seemed intuitively logical. But they saw no reason why a fast-spinning top should keep its axis vertical or even straighten it up, if it had not been vertical at the start. It was incomprehensible to them why, when the motion had slowed down, the axis should bend at an angle with the ground and move in such a way that the upper part of the top should describe a circle. And they could not tell at what speed the change took place.

A mystery is a challenge to inquisitive minds. Solving the mystery of the top became the boys' main preoccupation. They talked of nothing else, nothing else seemed to matter. The notions of mechanics that they had acquired from their textbooks were elementary and did not apply directly to the motion of tops. They refused to give up, and in the end Fermi arrived at a working theory of the gyroscope. The road he followed was laborious and roundabout. Had he been acquainted with two theorems well known to more advanced students, he would have saved himself much time and effort.

During the period from his brother's death to the end of high school, Fermi received guidance and advice from one of his father's colleagues. Enrico had fallen into the habit of meeting his father at his office and of walking home with him. Often Ingegner Amidei joined them. An effusive man of enthusiastic disposition, Amidei soon became impressed by the boy's clear thinking, his mathematical ability, and his interest in science. In a teasing mood, at first, he gave Fermi a few problems to work on, stating that they were certainly above his level and that he did not expect him to solve them.

But the boy did. He asked for harder ones and still succeeded. They were now problems that Ingegner Amidei had not been able to work out himself. The older man's interest in his young friend turned into admiration. He lent Enrico the few books he owned, one after the other, in a logical order, to build a sound foundation of mathematical principles and a basic knowledge of physics. Fermi, on his side, implemented the books borrowed from Amidei with those he purchased in a more random way at Campo dei Fiori.

Thus the idea of becoming a physicist, nursed and tended with care by his older friend, took root in Enrico. When he reached the end of high school, it was again Ingegner Amidei who gave him sound advice. There was in Pisa, he said, a little-known institution for outstanding students of letters and sciences, the Reale Scuola Normale Superiore. Enrico should apply for a fellowship there. There was little doubt that he would obtain it.

Fermi's parents were uncertain. It was not customary for a boy to study away from his family when there was a good university in the city in which they lived. They yielded to Amidei's insistence, in the end, and Enrico applied for the fellowship.

The examination that he was required to take provided him his first chance to make himself known to members of the academic world. He was asked to write a paper on vibrating strings. He poured into this work as much of his erudition as he possibly could. A puzzled examiner, a professor at the School of Engineering in Rome, incapable of explaining so much knowledge in a young man of his age, called him to his office for an informal conversation, at the end of which Fermi learned that he was "exceptional."

(3)

THE TIMES BEFORE WE MET—*Continued*

Fermi left Rome for Pisa early in November, 1918. He was seventeen years old. The first World War was coming to an end. Italy's traditional enemies, Germany and Austria, had been defeated. Trento and Trieste, the two cities for which the Italians had fought and had left six hundred thousand dead on the battlefields, had been liberated from the Austrian "yoke." A long-lasting peace was in sight. Young men would have to fight no more, and to them the outlook for the future was as rosy as it would ever be.

Fermi left home with a light heart and great expectations. They were not going to be thwarted. Perhaps because other students shared the feeling of relief, the shedding of insecurity, perhaps because Pisa was a small university city still carrying on Middle Ages traditions of gay student life, perhaps because he had left behind the gloomy atmosphere that had filled his home since his brother's death, perhaps for other imponderable reasons, the four years at the Scuola Normale were to be Fermi's most happy and lively.

The Reale Scuola Normale of Pisa had been established by Napoleon in 1810 as the Italian counterpart of the Ecole Normale Supérieure of Paris. Both institutions were meant to attract and develop young talent, a purpose that both attained beyond any doubt.

The Pisan school provided free board, lodgings, and some special lectures to "normalists," who otherwise were enrolled as regular students at the University of Pisa. Dormitories and classrooms were and still are housed in a sixteenth-century palace, one of those palaces to which proportion of lines gives a light and deli-

Giulio, Enrico (aged four), and Maria Fermi

Enrico Fermi at Sixteen

Emilio Segré, Enrico Persico, and Enrico Fermi (1927)

cate appearance despite the actual bulk and size. The cell-like bare-ness of the students' rooms was in contrast with the opulence of the building's front, finely but profusely decorated by Giorgio Vasari.

In Fermi's time there was no heating in the Scuola Normale, and winters were colder in Pisa than in Rome. But Fermi did not need to sit on his hands and turn pages with his tongue, as he had done at home, because each "normalist" was given a *scaldino,* a handled crock with smoldering charcoals and ashes. If kept in one's lap, it gave warmth to one's hands and stomach.

If in winter one had to fight against cold, in summer one fought against mosquitoes. Then elastic garters came in handy and could be shot at the flying insects, a sport at which Fermi claims to have been proficient to the point of becoming one of the best mosquito killers in the school.

Fermi's study hours were not long. He already knew a great deal of what was taught, and he easily remembered whatever new notions were exposed in the classrooms. He had much time left for the sort of fun peculiar to students' life in small university centers: fights with pails of water on the roofs of Pisa to protect two young damsels' honor, which had never been in danger; make-believe duels for reasons that were unknown both to challengers and to challenged, with consequent delving into the Chivalry Code: a frantic and successful campaign to elect the least attractive girl for May Queen, to her own embarrassment. I do not believe Fermi would have given himself so thoroughly to this kind of life if he had not been dragged into it and held fast there by a new friend, Franco Rasetti.

Rasetti, a first-year student of physics like Fermi, was not a usual person; his main interest was directed to that part of the world which is not made of human beings. He was a born naturalist. When a child of four, if given a pair of scissors and some colored paper with no drawings or marks on it, he would cut out insects, mantes, ladybugs, cockroaches, and butterflies, with so great a resemblance to real ones that nobody could have doubts about what each of them represented. As an adult he knew the names of some fifteen thousand fossils and about as many plants, a claim that nobody has ever been able to challenge. At fifty he would still run up the steep side of a mountain in pursuit of a butterfly—and catch it. Rasetti liked biology. He chose physics instead, because it was not easy for him to understand it and he wanted to prove to himself that he could overcome any difficulty.

Intelligence gave him no satisfaction; only a sense of futility, a spiritual restlessness, that made him seek excitement. He had organized a group of students among whom Fermi was prominent in an "Anti-Neighbors Society." The society's single aim was to pester people. The tricks they played ranged from placing a pan of water on a door left ajar, which would give a shower to the first person going through, to exploding a stink-bomb in a classroom during a solemn lecture. For the latter prank Rasetti and Fermi, who had built the bomb, risked being expelled forever from the university. They were saved by their teacher of experimental physics, Professor Luigi Puccianti, a tolerant man with keen judgment, who stressed their scholarly achievements at an especially convoked disciplinary meeting of the faculty.

Of one particular joke they were inordinately proud. Members of the "Anti-Neighbors Society" were requested at all times to carry a small padlock in their pockets, preferably painted a conspicuous yellow or red. They operated in pairs: while one of them entertained the predestined victim with pleasant or learned conversation according to the case, the other deftly slipped the lock shackle through two opposite buttonholes of his suit or topcoat, and—snap! the poor devil was trapped, and no amount of pleading would free him.

To be a member of the society was no safeguard against locks. On a certain spring morning Fermi, who always awoke very early, found himself all clothed when other students were still sound asleep. He quietly walked out of the front door and down the monumental stairway at the front of the Scuola Normale. In front of Rasetti's house he produced from his pockets two screwing eyelets, fastened one in the front door, the other in the doorway, and snapped a padlock through them. A little later a crowd of students gathered under Rasetti's window, shouted to him to come out, and took great delight in his fury at finding himself a prisoner in his own home.

On Sundays Fermi and Rasetti took long excursions on the Alpi Apuane, that region of the Apennines which is north of Pisa. Rasetti, as lively as a wire spring and as agile as a gazelle, rushed up the sides of the mountains. Only Fermi, despite his short legs, had enough wind to keep pace with him. Often, coming back late in the evening from these excursions, Rasetti asked Fermi to his home. He was an only son and very much attached to his mother, a small woman with more spirit than body, who had encouraged and directed his naturalistic trends and now looked at him in wonder as if surprised to have produced him. She liked Franco's new friend and invited him often for a meal.

The excellent dinners that Fermi ate at the Rasetti's interrupted the monotony of the food at the Scuola Normale, which consisted largely of dried cod. Those were the years immediately after the first World War, and many edibles had disappeared from the markets or had become excessively expensive. I remember eating amazing quantities of canned salmon during that period. It was dry cod at the Scuola Normale, and most pupils grumbled. Not Fermi. He was of simple tastes, and, moreover, he thought that complaining was an idle form of expression directed to no purpose.

In the stories of Pisa that I later heard there was seldom any mention of study. But the shadow of Pisa's greatest son must have loomed in that city and inspired young physicists. Galileo had experimented on falling bodies from the top of the leaning tower past which students in Fermi's times walked daily; the lamp whose oscillations had suggested the laws of the pendulum to Galileo was still hanging from the same ceiling in the old cathedral.

Perhaps Fermi and Rasetti breathed physics with the air in Pisa; certainly they learned much. Their physics teacher could do little more for them than give them a free hand in his laboratory. Professor Luigi Puccianti was a learned man, well versed in literature, who would have been more successful as a scholar of humanities than as an experimental physicist. He had intelligence and a critical mind, but he lacked drive. He had achieved some research in the past, but since then he had done little more than teach classes and wander around a laboratory, where disorder, rust, dirt, and cobwebs prevented him from working when he still felt so inclined. Soon his two pupils knew much more physics than he. He was aware of this fact, and he asked Fermi to teach him theoretical physics, "because," he said, "I am an ass, but you are a lucid thinker and I can always understand what you explain." Fermi, who was never hampered by false modesty, readily agreed and held a course of lessons on Einstein's relativity for his teacher.

In July, 1922, Fermi received his degree of Doctor in Physics. His thesis consisted of a paper on experimental work with Roentgen rays (X-rays). He also gave an oral dissertation. The latter was open to the public, and friends gathered in the Aula Magna expecting to witness Fermi's triumph. They were to be disappointed.

The eleven examiners in black togas and square-topped hats were sitting in solemn dignity behind a long table. Fermi, also in a black toga, stood in front of them, and he started talking with cool, deliberate assurance. As he proceeded, some of the examiners repressed yawns, some sent their brows up in wonder, others relaxed and did not try to follow. Clearly, Fermi's erudition was above their comprehension. Fermi received his degree *magna cum laude*. But none of the examiners shook hands with him or congratulated him, and the customary honor of having the thesis published by the university was not conferred upon him.

Fermi went back to Rome and to his family.

On September 29 he became of age. Twenty-nine days later a great historical event took place.

I must qualify the meaning of the word "historical," or it would not correspond to my personal feelings about the event. Custom and the dictionary have attached an objective and dispassionate connotation to "historical." History, we are given to understand, is the complex of past experiences that have been completely lived and that are external to our existence. The living and history do not mingle. Issues and facts become history when they cease to be vital and to affect us in any way. Our days belong to us, not to history.

These statements I accepted so long as I was young. But as years went by, I came to realize that they were wrong. We, the living, are part of history and cannot escape its consequences. History has shaped our lives. One single occurrence, one single day, may have sharply changed our fate.

Whenever in my memory I retravel upstream the course of my existence in search of the source from which all happenings of significance to me, to my family, and to my friends have flowed with predestined inevitability, I always reach the same point: the March on Rome.

I could not realize its importance at the time it took place. Its consequences may have been foreseen by a few enlightened thinkers and by prophets. I was fifteen, and to me that day was one marked by some strange occurrences but was not essentially different from others: its basic routine was not disturbed.

I remember that October day vividly. Because I am easily impressed by the incongruous, a visual incongruity is most prominent in my memory: the handles of two short daggers sticking out of the fur cuffs of my sister's and my winter coats. There must have been cold days before, because our winter coats were out of moth balls and hanging in the hall. But on the twenty-eighth the weather must have been mild, and we left the heavy coats at home.

The day started like any other day, without warning or foreboding. My sister Anna and I, who were steadily together and formed a pair, kissed our parents goodbye in their room—at eight in the morning they were accustomed to have breakfast in bed—and left for school. The girl across the street joined us, and the three of us walked as we always did, books under arms and at a slow pace. Stores were being opened; streetcars sped by with their usual clusters of people hanging outside their doors; students walked to their schools; mothers and maids dragged younger children by the hand.

We entered our school, the Ginnasio-Liceo Torquato Tasso, we put on our black uniforms in the vestibule, and we went to class. At about eleven the old janitor walked into our classroom, dragging his tired feet behind him, and handed a sheet of paper to the teacher. That, we knew, was a circular from the principal. The teacher started reading it.

We could not accord to that particular teacher the respect we owed him, because of his clownish face: a huge purple nose under a pair of large blue glasses. They hid one good eye and one that had been blinded with a pistol shot by a flunked student.

The teacher finished reading the circular and his face paled.

"You are dismissed," he said, and his sharp tone awed us. "Go directly to your homes. Don't linger in the streets. Grave happenings are taking place. Columns of Fascists have entered Rome from all directions, through all its gates. The cabinet has proclaimed a state of siege."

We made for home, my sister and I. Rome looked different now. The Via Sicilia was filled with the boys and girls pouring out of our school and with the idle onlookers who crowded the sidewalks. Soldiers and *carabinieri* stood at each corner. We reached the Flavian walls at Porta Salaria, one of the ancient gates. It was closed to traffic by barbed-wire entanglements, the same that had been used in World War I and the like of which I had never seen. Sentinels with bayonets fastened on their rifle barrels stood guard by them. The Piazzale di Porta Salaria, on the other side of the gate, was less noisy than usual because the streetcars were not running. The few cars and horse carriages that tried to cross the square were stopped by soldiers, their drivers questioned, then motioned to proceed. A truck came from the direction of Villa Borghese, loaded with youths in black shirts, shouting loudly, and the Italian flag that they carried waved in the air.

My sister Anna and I went on. We walked close together to draw reassurance from each other. We reached Piazza Principe di Napoli, the market place. The benches with fish and meat, fruit and vegetables, stood as they had earlier in the morning, under the large colorful umbrellas. Vendors and shoppers talked to each other in animation. On their faces were astonishment and uncertainty, expectation and concern.

Farther on, the quiet section where we lived among small villas inclosed in green gardens was not so quiet as other days. Even in our short, unimportant street there was commotion. As we turned the corner into our street, three young men in black shirts and rough clothing came out of our garden gate with big, rushing strides and excited gestures.

We went inside the house and met our parents, waiting anxiously for us in the doorway. We walked into the hall. Then my eyes fell on two objects that did not belong there, half-hidden behind the long rabbit fur, in two cuffs of our coats.

"What's that?" I asked. And as they followed my glance, my parents and my sister repeated: "What's that?"

Then I extracted from our fur cuffs two short, wide-bladed knives.

"Those boys!" my father exclaimed with surprise and condescension, "Those boys! They cheated. They did not surrender all their arms to the police. Then they got cold feet and hid their knives in your cuffs!"

The boys, he said, were Fascists from somewhere near Rome. Like many others, they had come in a small group, without waiting for the formal order to march. In their villages they had grabbed whatever arms they could lay their hands on. They had arrived in Rome ahead of the main columns. While wandering around, uncertain of their role, disoriented, they had read the cabinet's proclamation of the state of siege, which was being hurriedly plastered on all walls. The civil organization of the city, they read, was being turned over to the military, and the army would receive instructions to resist Fascist invasion.

The three young men could not know that the army, in sympathy with the Fascist movement, would do nothing while waiting for more specific orders; that these could not be issued as long as the king would not sign the decree of the state of siege; and that the king would refuse his signature with unexpected determination.

They became scared. They thought they might be arrested, sentenced to harsh punishment under military law for illegal possession of arms. They sought sanctuary in our home. My father, paternal toward them but used to exacting obedience, had first lectured them on law and order, then had escorted them to the nearest police station to surrender their firearms. That they could conceal knives had not occurred to him.

Enrico Fermi, as I learned much later, had spent that morning at the physics building with Professor Orso Mario Corbino, the head of the department. On his return from Pisa with a degree in his pocket Fermi had gone to Corbino in order to discuss with him the possibilities for the future. It is always with a certain hesitation and apprehension that a young man with no official status visits for the first time one who has an established position and reputation. Professor Corbino was a member of the Italian senate and a prominent figure in government. He had been a minister once and was to be one again. To his relief, Fermi had found him affable, well versed in modern physics, ready to listen, and easy to talk to. On his side

Corbino, impressed by the younger man's knowledge, had encouraged him to come back for frequent talks.

Thus on the morning of October 28 Fermi was in Corbino's office. This time they did not talk of physics. Nor did they make plans for future work. Corbino was preoccupied with the political situation. He did not like the profession of violence made by the Fascist leader, Mussolini. That young man was tough, ruthless. The columns that on his order were entering Rome constituted a threat and a danger to the country.

"But," he said, "the cabinet's decree of state of siege is no solution. It can bring nothing good. If the king signs it, we may have a civil war. The army will be ordered to fight, and if they obey, if they don't pass to the Fascists' side, there is little doubt of the outcome: Fascists have no arms; there will be a massacre. What a pity! so many young men will die who were only in search of an ideal to worship and who found none better than fascism."

"You put in doubt the king's signature. Do you think he may go against his cabinet? He has never been known to take the lead but has always followed his ministers."

Corbino pondered a minute.

"Yes," he said, "I think there is a chance that the king may not sign the decree. He is a man of courage."

"Then there is still a hope . . . ," Fermi said.

"A hope? Of what? Not of salvation. If the king doesn't sign, we are certainly going to have a Fascist dictatorship under Mussolini."

That evening Fermi related this conversation to his family. The king's refusal to sign had been made public. Fermi had absolute faith in Corbino's soundness of judgment and clairvoyance. There was not the least doubt in his mind that a dictatorship was impending.

"Which means," he concluded with the cool detachment that he retained when uttering the most ominous predictions, "that young people like me will have to emigrate."

"Emigrate?" his sister repeated with anxious tenseness. Maria was preparing to teach Latin and Greek in public schools. A scholar, she had plunged into the study of the old texts. She strongly felt the classical traditions, the claims of ancestry, the continuity of generations descending from that ancestry, the acceptance of the past into the present, all the constraining ties that bound her to the country where she was born. To emigrate meant to tear those ties abruptly, to go against a nationalism that was no less compelling for not having been formulated by the family, for having been allowed to smolder quietly under their professed liberalism.

"Emigrate?" she asked. "Where to?"

Fermi shrugged his shoulders:

"Somewhere . . . the world is large."

But he stayed. And the fact that sixteen years later almost to the day he left Italy for the United States does not make a prophet of him.

The following winter Fermi went to Göttingen in Germany with a fellowship from the Italian Ministry of Public Instruction, to study with the well-known physicist, Max Born.

In Göttingen for the first time in his life Fermi savored the taste of wealth. The inflation in Germany was then spiraling upward at tremendous speed. Fermi exchanged the weekly instalments of his fellowship at an increasingly advantageous rate. Although occasionally he felt the pangs of one who, having concluded a deal which appeared extremely profitable in the morning, realizes suddenly that he could have doubled his gain just by waiting until the evening, still he was rich. And he experienced the pleasures that riches can give. By the end of his seven months' stay in Göttingen, Fermi invested his savings in a brand-new bicycle for himself.

However, intellectual security did not come together with financial security for Fermi. In Germany his old shyness returned and hampered his social relations. The language did not bother him. He could speak some German before he went to Göttingen, and soon he had mastered it well. Still he could never shed the feeling that he was a foreigner and did not belong in the group of men around Professor Born.

Born himself was kind and hospitable. But he did not guess that the young man from Rome, for all his apparent self-reliance, was at the very moment going through that stage of life which most young people cannot avoid. Fermi was groping in uncertainty and seeking reassurance. He was hoping for a pat on the back from Professor Max Born.

Fermi knew himself to be held in good esteem by scientists in Italy. He also knew that in the kingdom of the blind the one-eyed man is king. How could he tell whether he had one eye or two as a physicist? What was the measure of his abilities by absolute standards? Would he be capable of competing with young scholars like those around Professor Born, so keen, so learned, of whom Werner Heisenberg was one? The seven months in Göttingen dragged on inconclusively, with little profit, adding to the uncertainty.

Fermi was to receive the wished-for pat on the back more than a year later, from Professor Ehrenfest, in Leiden. Meanwhile, his fellowship had expired, and he went back to Rome, where he was to teach an elementary course of mathematics at the university. And there he was in the spring of 1924, when I first met him.

(4)

BIRTH OF A SCHOOL

By the fall of 1926 Fermi was settled in Rome for good. He had stored away his Tyrolean jacket and his knickerbockers, and he had donned a too tight suit, the same hazelnut color as his tanned face. He was full professor of physics at the University of Rome.

To a mere student like me, a professor seemed to wear an overwhelming halo of importance and solemnity. After the teasing companionship of the previous summer I met "Professor Fermi" with mixed feelings. But the young physicist who could inspire respect in his older colleagues showed a remarkable ability to put himself on the level of the young, and I found that I could still talk to him without restraint. Often on Sundays I joined him and his group for a hike in the country or a stroll in Villa Borghese, the main park of Rome. Our companionship did not break up.

That same fall Fermi introduced to us his friend Franco Rasetti, an elongated man with thin hair, a determined chin, and a steady gaze that went through people. What impressed me most upon meeting him was that he spoke in exactly the same fashion as Fermi, with a queer drawl, a slow singsong mixed with a strong Tuscan accent. It was clearly a case of a contagious disease, but who had come down with it first I could not ascertain. Fermi's sister, who strongly resembled her brother but for her fair complexion, had contracted the disease in a lesser degree. This fact proved nothing. It was possible that the bug had come to Maria from Rasetti through her brother and had lost some of its virulence on the way.

Rasetti's laughter and Fermi's were also similar, loud and prolonged outbursts of mirth at the least provocation. But Fermi's was of lower pitch. Rasetti's, harsh and unrestrained, was a true cackling. It was not in line with the meticulous precision of all his actions, which made him often bend his head to inspect his suit lapels and remove a speck of dust from them with the tips of his deft fingers.

Girls were attracted by Franco Rasetti, although, or perhaps because, he remained unaffected. He looked them over with dispassionate detachment, bending his head to one side for a better view, with narrowed eyes behind his glasses. He examined them, he dissected them with his piercing look, as if they were rare butterflies or strange plants, and all along there was a smile of amusement on his mouth, tinged with a bit of mockery. He could talk of almost anything, from Buddhism to English novels for girls —*Jane Eyre* was the subject of my first conversation with him. But he was a physicist, as Fermi told us, and had come from Florence to teach physics at the University of Rome.

It could be no coincidence that Fermi and he had come to Rome at the same time. Rome is the most desirable residence in Italy. It was "Caput orbis terrarum," the head of the earth, to our Latin ancestors. To our generation it is the Eternal City, the most wonderful in the world. Not often could young men of twenty-five bypass the line of hopeful professors at other universities waiting for their chance to come to Rome. Actually, it was not a coincidence. There was a reason for it, the fact that for some time Senator Orso Mario Corbino, the head of the physics department in Rome, had been dreaming ambitious dreams for his school.

Corbino, a short man sparkling with vitality, was born in Sicily and shared the qualities common to most Sicilians: mental acuteness, keenness of judgment, and combativeness. That same combativeness that, misguided, had produced the Sicilian Mafia had become in Corbino an unlimited driving force, coupled with the power of attaining any goal he aimed at. A self-made man, he had reached a prominent and influential position in politics. Wise with the wisdom of this world, he achieved the miracle of being a minister in Mussolini's cabinet without ever joining the Fascist party.

Senator Corbino was aware that physics had come to a standstill in Italy. Physicists of his generation, lulled in the old glories of Galileo and Volta, made no effort to maintain the tradition. Except for mathematical physicists, who in reality were mathematicians, Corbino himself had been the only physicist of some stature during the first quarter of our century. Now, in his mature age, he·divided most of his time between his political responsibilities and his advisory offices with industries.

Though he had abandoned active research almost entirely, still his mind and heart were in physics. He had bold visions of a great school capable of important achievements, that would spring up in Rome and gain recognition the world over.

A vision was a challenge to Corbino. His dreams must come true. The first step in building a good school is to gather good men. For a while Corbino had been playing with the idea of calling Fermi to Rome and giving him a permanent position. In the academic year 1923–24 Fermi had been an instructor in Rome, teaching mathematics for chemists and science students, a course traditionally held by physicists in the physics building. In 1926 Corbino wanted him to come for good.

Fermi was teaching mechanics and mathematics in Florence, as *incaricato*, a position carrying no tenure or pension privileges. The year before, he had tried to obtain a chair of mathematical physics at the University of Cagliari, in Sardinia. According to Italian law, a university may fill a vacant chair in a certain subject by holding a *concorso*, or competition, among applicants. There is no formal examination, and a committee of professors from several universities selects applicants on the basis of their curriculums and publications. The three best candidates become qualified to occupy chairs in that subject; the first obtains the position at the university holding the competition, the second and third may fill vacancies elsewhere, if and when they occur.

Cagliari is a small, remote city. Sardinia provides excellent material for folklore students but lacks the cultural atmosphere and the comforts of more developed regions. A position at the University of Cagliari is considered a steppingstone to better residences.

When Cagliari announced its *concorso*, Fermi decided to apply. He sent his qualifications with confidence, because, although the youngest candidate, he knew that he had an established reputation in academic circles both in Italy and in foreign countries. He had already published some thirty papers, part on experimental work, the majority on theoretical studies mainly in the field of relativistic theories. Unluckily for him, the Italian mathematicians of that time were divided into two factions: those who had mastered Einstein's relativity and who accepted it and those who "did not believe in it."

The examiners in Cagliari were also divided, three of them being anti-Einstein and only two, Professors Levi-Civita and Volterra, belonging to the pro-Einstein faction. These two men, both of international fame, both teaching in Rome, had broad views and were deeply interested in new scientific developments. They had had numerous opportunities of meeting and appreciating Fermi, his lucid thinking, his prompt grasp of any concept, his ability to reduce problems to their essential elements. Fermi was their candidate.

The other three examiners favored another man: Giovanni Giorgi, a middle-aged engineer, who had received his degree eight years before Fermi was born and had achieved a certain reputation with his M.K.S. system of units. This was a different choice of the fundamental units of physics, presenting some practical advantages. Now his supporters stressed his "greater maturity," "the entity of his production," "his speculative and philosophical intellect." He placed first and was appointed to the chair at Cagliari.

There were no other vacant chairs in mathematical physics, and Fermi stayed in Florence. In 1926, when Corbino thought of calling him to Rome, Fermi was still available for a permanent position. He had also better claims to one, for in the year gone by his reputation had grown. He had published a statistical theory "On the Quantization of a Perfect Monatomic Gas," a work that found its place among the most representative in the evolution of theoretical physics.

For a few years Fermi had been interested in statistical questions, the behavior of molecules, atoms, and electrons; the energy distribution in radiation emission. He had given much thought to the behavior of a perfect, hypothetical gas. The precise law which such a gas obeys had baffled him for some time. Some factor that would bring full comprehension was missing, and he could not figure out what it was.

Scientific problems seldom stand alone, but their solutions often interlock with one another. For over ten years theoretical studies had been directed to a full description of the atom and of the atomic laws that would bring order and understanding to the confused stockpile of experimental data. It was a boom period for atomic physics; and new theories, new principles, and viewpoints shedding new lights on old concepts were worked out and published in rapid succession.

In 1925 the Austrian-born physicist, Wolfgang Pauli, while studying the energies of the atomic electrons rotating around the nucleus, discovered his principle of exclusion, which, without the accuracy of scientific language, reads: On each of the orbits around a nucleus there can be only one electron. At once Fermi extended this principle to a perfect gas.

In those days Fermi had much time for speculation. The physics laboratories of the University of Florence were in Arcetri, on the famous hill where Galileo had made his home during his last years and where he had died. While teaching at the University of Florence Fermi lived in Arcetri. Led by his friend Rasetti, he spent long hours chasing geckos, small lizards, perfectly harmless despite popular belief that they house evil spirits in their bodies. Fermi and Rasetti were going to let their captured geckos loose in the dining hall for the simple pleasure of scaring the peasant girls who waited on tables.

The two friends lay for hours stretched on their stomachs in the grass, in perfect stillness, each holding a long glass rod with a brief silk lasso at its end. During the patient vigil Rasetti observed the small world under his eyes: a tender blade of growing grass, a busy ant scurrying by with a bit of straw in its mouth, the play of a ray of light on his glass rod. Fermi, who is no naturalist, was not interested in that small world. While he watched the ground, ready to pull his lasso should a gecko appear, he let his mind wander. His subconscious worked on Pauli's principle and on the theory of a perfect gas. From subconscious depths came the missing factor Fermi had long sought: no two atoms of a gas can move with exactly the same velocity or, as physicists say, there can be only one atom in each of the quantum states possible for the atoms of a perfect monatomic gas. This principle permitted Fermi to work out a complete calculation of the gas behavior, known as "Fermi's statistics." This, in its turn, was later used both by Fermi and by other physicists to explain a number of phenomena, including the thermal and electric conductivity of metals.

Fermi's statistical paper was just out when Corbino advanced the proposal of establishing a new chair in theoretical physics. The faculty of science received the proposal with favor, and Corbino's prominent personality was instrumental in securing the necessary legislative decree from the Minister of Education. The *concorso* for the new chair was called in the summer of 1926, and Fermi placed first. The second on the list of three winning candidates was Fermi's old friend, Enrico Persico, who went to teach in Florence. The third winner, Aldo Pontremoli, disappeared in the Arctic Sea two years later, while he was a member of an unfortunate Italian expedition to the North Pole. Fermi obtained the position in Rome and started teaching on the following October.

One man alone does not make a school. On the lookout for more talent, Corbino became aware of Rasetti and called him to Rome as *aiuto*, a position requiring no *concorso*. Corbino had now a staff of two eager teachers, but physics students were few and of poor quality. The best young people with scientific aptitudes were attracted by the excellent but arduous School of Engineering. When a student who had set too high aims for his abilities realized that he would never master the overloaded curriculum, he would transfer to physics, where he received credit for his completed courses. For several years this had been the main source of physics students, a fact that clearly explains why their level was so low. Corbino resolved to change this condition.

I was then a second-year student of general science and was following Corbino's course of electricity for engineers. Students of general science, who were not numerous (six girls, two boys, and two priests in my year), found themselves at a disadvantage. No special courses were offered them, and they had to attend those for students of medicine, engineering, and chemistry.

Before the Wedding . . .

... And at the Wedding (1928) : The Naval Officer Is My Father, to His Left Is Senator Corbino, and My Mother Is Second from the Right

The present University City had not yet been built. It was being planned as one of those "realizations of the Fascist regime," of which the Fascist press bragged daily and loudly. Meanwhile, class-rooms were scattered to the four corners of Rome. Law, literature, and social sciences were housed in the old university building "La Sapienza" (this word means "knowledge") designed by Michelangelo; but the various branches of science were not so lucky. I used to spend much time traveling by streetcar from the Sapienza, where some forgotten scientific courses were still held, to the zoo, where a small building left vacant by animals had been temporarily loaned to zoölogy; and from there to the opposite side of the city for my physics and chemistry classes.

Chemistry and physics were in two buildings that still stand on hilly grounds along Via Panisperna. Bamboos and palm trees grow along the graveled alleys that link the buildings. Two convents had been on those grounds before 1870. In that year the pope had lost his temporal state, Rome, to the conquering armies of the newly united Kingdom of Italy. The new government was liberal and progressive. The two convents with the grounds around them were assigned to physics and chemistry, respectively.

The chemists of that time welcomed their convent as well suited to their needs. They did little more than close the walks around the cloister with glass walls to transform them into laboratories, leaving the old stone well in the center, with pulley and water pail, as if in use. They brought no change to the exterior: lofty, heavy, and bleak, the chemistry building looked more like a prison than a hall of science.

The physicists, on the other hand, did not relish the idea of turning cells into classrooms. They decided to have their own convent demolished and a modern institute built in its stead. But the nuns who had lived in that place for centuries refused to leave. Neither threats nor enticement would persuade them. In the end the physicists dispatched a detachment of *bersaglieri* up the hilly alley of the convent. *Bersaglieri* were specially trained, swift-run-ning infantrymen with wide feathered hats. I have it from a good source that when the nuns saw the feathers from afar, they rushed to pack.

The new building was of sober architecture. It was planned with foresightedness and largeness of means. It was well equipped by European standards when I was in the university.

Corbino held his classes in a wide room with rows of benches that came down toward the front from a considerable height near the back wall. The teacher's desk was on a raised platform. Corbino, short and chubby, hardly emerged from behind it. When he climbed the platform with little jumpy steps or when he jerkily ran to the blackboard and lifted his body from the floor in the effort to write high enough, he cut a clownish figure. But when he started talking, the room became still, the attention of the class centered on the shiny head. The little man became impressive. Electricity, a subject which I loathed, became temporarily enjoyable.

One morning Senator Corbino made an announcement in class: he was looking, he said, for two or three outstanding students who would like to transfer from engineering to physics. He could assure them that this was an extraordinary opportunity: modern physics was the most promising field of research. The staff of physicists at the university had just been enlarged and was ready to give personal attention and excellent training to a few students. The outlook for future positions at the end of the training period was also very good, although on this point he could not give a formal guaranty. Corbino's eyes were two black dots of concentrated wit as he scanned the rows of benches and repeated that he wanted only the best available students, worthy of the time and effort that would be spent on them.

Corbino's plea reaped a single student, Edoardo Amaldi. He had come to Rome only two years before, when his father, formerly a professor of mathematics at the University of Padua, had joined the Roman university. Edoardo was in the second year of the School of Engineering and attended Corbino's course in electricity. Eighteen years old, he looked much younger: under a heavy mass of curly chestnut hair his face was rosy and soft; his lips were full and the color of coral.

Amaldi and I were good friends, and he used to come to my parents' home when we had gatherings of young people. I shall always remember a party at which we played a certain game of our invention, "silent movies." A group of us would silently act one of the best-known films, while one person would recite the captions and another would emit an uninterrupted buzzing sound to reproduce the noise of a projector.

Fermi had also come to our party and taken the lead, as was his habit: he was film director. He assigned the roles to us, and we usually accepted them with no questions. But when he asked my sister Anna to be Greta Garbo, she flatly refused. Thin and tall, with a dreaming expression, Anna was the natural choice for that role. But she was also bashful, in a sense, and stubborn. Fermi had no ascendancy over her, because, being an artist, she held the whole class of scientists in little consideration. For once, Fermi had found a match for his determination. Gina and Cornelia had their roles already. No more girls were available, and we almost gave up the film. But Fermi proved the extent of his resourcefulness by turning to Amaldi:

"Show the girls that we men have no false shyness. Take the role of Greta Garbo!"

Edoardo, infected with Fermi's mood, accepted with good grace. Soon he appeared on the makeshift stage in a sky-blue velvet dress that he had borrowed from me, with a low neck opening revealing a vast whiteness worthy of the most celebrated stars.

After this performance I could not take Amaldi too seriously. When he mentioned his intention of following Corbino's advice and transferring to physics, I teased him for the good opinion in which he held himself. But Edoardo, like the rest of the Roman group, was not bothered by inferiority complexes. After a not too long period of deliberation he came to the conclusion that he was the very student to fill Corbino's requirements and was soon accepted into the expanding family of physicists and physicists-to-be.

One of the places where I used to meet the young people from the physics building was Professor Castelnuovo's apartment. He and his family were at home to all their friends every Saturday evening after dinner, and I started to attend their gatherings the year before Fermi came to Rome. I had then learned that one of my teachers was a steady visitor at the Castelnuovos' Saturday evenings.

In 1925–26 that same Enrico Persico who had been Fermi's

friend in the Liceo was teaching the first-year course of mathematics for chemists and general science students, which I was required to take. The relations between teachers and students were formal and impersonal. They saw each other only in crowded classrooms, and they were given no opportunity for more human relationship.

Prompted by the desire to learn how a teacher behaves in social company and by the pleasurable prospect of later bragging to my school friends about my encounters with the blond mentor, I persuaded my parents to accept a standing invitation and pay a call on the Castelnuovos.

In the Castelnuovos' small parlor ten or twelve middle-aged people were sitting in a circle on green plush chairs. The chairs were low and narrow; the walls were high and unadorned. Dim bulbs hidden among innumerable glass drops in the ceiling lamp shed a frugal light over the gray heads.

Like my father, most men wore beards; like my mother, most women were clad in black. The men were the greatest Italian mathematicians of that time. Several of them had attained international reputation: Volterra, Levi-Civita, Enriques. All were eminent in Italy. Mathematics had attracted some of the ablest men, and the group of mathematicians was outstanding for their talent, their achievements, and their moral soundness.

Saturday after Saturday they gathered at the Castelnuovos with their wives and children, to spend a few hours in informal chatter among congenial friends. They talked of the latest happenings in the faculty of science: of births and deaths; of marriages and flirtations; of faculty policies; of new discoveries and theories; of the rising stars in physics.

I was a first-year student at the university when I entered the Castelnuovos' parlor with my parents, and I felt deeply awed by the great men. It was almost a relief to be told that I could go into the dining-room.

A group of young men and girls was sitting around the dinner table covered with a green rug. I took my place between Persico and a rosy, cherubic youth, Edoardo Amaldi, the same who in the following year was to answer Corbino's bid for good students. My mathematics teacher graciously acknowledged my presence by shaking hands with me. He was shy and restrained for the rest of the evening, although we talked lightly and played games.

At ten, Daisy, the maid, placed a tray with cookies and fruit juice on the marble-topped cupboard of dark wood. She was old and motherly, and she treated the young people with the familiarity of one who had seen them in their cradles.

"Don't spill the juice," she cautioned Gina. Then, with a pat of her plump hand on her white apron and a "Good night to everybody. Have a nice time," she retired, swinging her ample flanks as she walked.

I found the group entertaining, and often that same year I dragged along my unwilling but inseparable sister Anna, whose artistic temperament made her look down on the positive sciences.

"I can't imagine what you see in those people," she told me after a few Saturdays. "They are all so uninspiring." Then she added disdainfully: "All logarithms!" That nickname was accepted in my family.

Next fall the group of logarithms underwent some changes: Persico left Rome, for he had obtained the chair of theoretical physics in Florence; Fermi and Rasetti came to sit around the Castelnuovos' dining-room table. Now and then a member of the group took along a new friend. Thus on a Saturday evening Emilio Segré made his appearance among us. He was a student of engineering a couple of years further on in his studies than Amaldi and I. He was not taking the course on electricity when Corbino made his appeal for good students, and he did not know about it. But after he became acquainted with Fermi and Rasetti at the Castelnuovos', he began giving favorable attention to the new school of physics.

Emilio Segré was a good judge of men. Although he disapproved of silly jokes and games; although he disliked to play "fleas," a game of Fermi's invention, consisting in making pennies jump on the table rug; although he could not understand Fermi's childish pride in always being the winner, still he realized at once that there was more in Fermi and Rasetti than met the eye.

Segré was not a person to draw rash conclusions or to be satisfied with those of other people. To form a fair and firsthand opinion of the two physicists, he attended the seminars that they held

43

at the university. He became more and more impressed that "in physics there were men who knew what they were talking about."

Segré was of a cautious nature. He considered the possibility of giving up engineering and transferring to physics. He talked of this to the brightest of his fellow-students, Ettore Majorana. He started reading physics books. In September, 1927, he followed Fermi and Rasetti to an international meeting of atomic physicists at Como. By the shores of that romantic lake, the greatest scientific stars had gathered.

"Who is the man with the soft look and the indistinct pronunciation?" Segré might ask his two mentors.

"That is Bohr."

"Bohr? Who is he?"

"Fantastic!" Rasetti would exclaim. "Haven't you ever heard of Bohr's atom?"

"What is Bohr's atom?"

Fermi would explain. Then Segré would ask about others and about their achievements. Compton, Lorentz, Planck: Compton's effect, Planck's constant. . . . Segre was learning a lot of physics. He liked it.

When school opened the next November, Emilio Segré was a fourth-year student of physics. Soon Ettore Majorana followed suit. The school of Rome dreamed of by Corbino was well under way.

In the years that followed, other men came to work in Rome, new students and postgraduates from Italian universities and from abroad, attracted by the name that the Roman school was acquiring. They came and went. But the first group stayed. Fermi, Rasetti, Segré, and Amaldi—Corbino called them "his boys"— were the nucleus of a team that worked for years in free and easy co-operation. The successful adjustments they made to one another's personalities produced strong ties of affection that grew steadily as they went through youth into maturity.

When the four came together, two teachers and two students, they were all young—there were seven years' difference between the oldest and the youngest. They shared a love for physical exercise, a swim in the near-by sea, a climb in the mountains, a long hike, and a game of tennis.

For Fermi and Rasetti, who had taken it up while they were both in Florence, tennis had precedence over all other activities. The day I was to take the comprehensive exam at the end of a two-year physics course, two of the three examiners were to be Fermi and Rasetti, a much dreaded binomial—"Fermi and Rasetti together flunk with no discrimination," I overheard two pale students mutter on an exam day. They had not yet been mellowed by age and had set as high standards for the students as they had for themselves.

I was lucky, however. Fermi and Rasetti were delayed by a tennis match and were replaced at the last moment by older, softer-hearted teachers.

Rasetti, Fermi, and Amaldi also shared a certain playfulness, a naïve love of jokes and silly acting that they brought into their serious work. Edoardo Amaldi's wife Ginestra well remembers the first time she was introduced to a teaching session in Fermi's room, when she was just Miss Giovene, a student of physics. A few years younger than Edoardo, Ginestra had entered the university after him. She followed the regular courses, and, in addition, she was asked to attend the informal meetings led mostly by Fermi, sometimes by Rasetti.

"Don't be scared," Fermi told Ginestra Giovene, "all we do here is to play a game. We call it the game of the two lire. Anybody can ask a question of anybody else. The person who does not give the right answer pays one lira. But if the one who asked the question cannot provide a satisfactory answer, then *he* pays two lire. As you can see, it's all very simple. Now let's start. Who has a question for Miss Giovene?"

Edoardo, who from his teachers had learned the technique of always talking half in jest, half in earnest, said he had a question ready, and one most suited to a woman.

"As you know, the boiling point of olive oil is higher than the melting point of tin. How can you explain that it is possible to fry in olive oil inside a tinned skillet?" (The best Italian skillets are made of tin-lined copper.)

Despite her trepidation Ginestra was able to figure out the correct answer:

"Oil does not boil, when frying. It's the water in the food that boils!"

All was not so simple as Fermi had said, and Ginestra had to learn more physics than the laws of cooking. But the teaching remained as informal as on that first day. Questions were asked as they came to mind, one leading to the other, not according to pre-established plan. Students and teachers pooled their efforts in solving their problems. Perhaps, Ginestra said, students learned so much from those sessions because Fermi, too, brought his problems to the others and worked at them on the blackboard, aloud, with chalk and voice, thus showing how a rational mind reasons, how accidental factors can be discarded and essential ones taken into light; how analogies with known facts help to clear the unknown.

Permanent proof of this lack of formality is an old desk that used to be Fermi's in Via Panisperna and which is now kept in the new physics building at the University City. It has a hole where Segré's fist once came down with a bang because the others talked out of turn and did not give him a chance to be heard. Segré was famous for his touchiness. It earned him the name of "Basilisk," and, like that fabulous serpent, he was said to throw incinerating flames from his eyes when his feelings were hurt.

Ettore Majorana went only a few times to the sessions in Fermi's room. Majorana was a genius, a prodigy in arithmetic, a portent of insight and thinking power, the most profound and critical mind at the physics building. Nobody bothered to use a slide rule or to write down numerical calculations if Majorana was available.

"Ettore, will you tell me the logarithm of 1,538?" one would ask. Or "How much is the square root of 243 by 578 cubed?"

Once he and Fermi had a race: Fermi worked with pencil, paper, and slide rule; Majorana with his mind and nothing else. They came out even.

But Majorana was a strange person, an introvert, shy of everyone. In the morning, while riding the streetcar to the university, he would set himself to thinking, his dark brows knit together. Often an idea came to his mind, the solution of a stubborn problem, or a theory that correlated experimental facts with one another. Then he would search his pockets for a pencil. A package of cigarettes provided paper, and on it he would scribble some

figures. Out of the streetcar he would walk into the physics building, still absorbed, with bent head, while a big black mop of un-
combed hair fell low on his eyes. He would look for Fermi and
Rasetti, and, package of cigarettes in hand, he would expound
his idea.

"Excellent! Write it down, Ettore, and have it published."

"Oh no!" Ettore shrank at the mere mention of a printed paper,
at the thought of letting strangers pry into his mind. "Oh, no. It's
child's play." He would smoke the last cigarette and throw wrapper
and figures into the wastepaper basket.

Majorana had thought out Heisenberg's theory of the nucleus
with neutrons and protons as building stones before Heisenberg
published it, but he never wrote it.

What other students learned from Fermi and Rasetti, what Fermi
had to learn gradually by teaching others, Majorana could learn
best alone in his room, where no human presence could disturb the
precarious balance of his soul. Only when he knew that unusual
and outstanding studies were to be pursued, would he come. The
quantum theory was one of these.

When Fermi undertook to expound the quantum theory in all
its extent and depth, the group found it hard to grasp and alien
to common ways of thinking. That both matter and energy con-
sist of bundles of waves is rather a dogma than a truth to be proved
by reason, they contended. It is a matter of faith. In matters of
faith the pope is infallible. In quantum theory Fermi is infallible.
Ergo Fermi is the pope. From then on, the "Pope" he was. This
appellative startled newcomers at first. But soon the news of Fermi's ascent to the dignity of the tiara spread into the international
world of young physicists.

Rasetti, who disclaimed full comprehension of quantum theory
but was still the best qualified to take over whenever Fermi was
absent, became the "Cardinal Vicar." Dark-eyed, Spanish-looking
Majorana was never satisfied with first mathematical proofs but
pursued each study further, burrowing deeper and deeper with his
keen insight, probing, questioning, pointing out errors. Ettore Majorana was named the "Great Inquisitor."

Then one day Persico appeared at the physics building. He
had come from Turin, the bearer of sad tidings. In Turin, he related,

no one—but no one at all—believed in quantum theory. In Turin everybody held it to be contrary to established truths.

The Pope was concerned. On the spot he named Persico "Cardinal for *Propaganda Fide*" and instructed him to preach the gospel to the heathen.

Persico's mission was most successful. After not too long a time he sent a detailed report to the Pope, in the form of a long narrative poem. He told his vicissitudes in the land of unbelievers and ended with the description of his teachings to the converted heathen:

> They believe with true, deep faith
> To which reason has now bowed
> That the light is wave and matter,
> Matter, wave the electron is.
>
> This is one of many dogmas
> That he teaches to the Heathen
> With examples to comfort them
> Which he draws from Holy Gospels.

The "Holy Gospels" was a book on quantum mechanics written by Persico and later translated into English.

(5)

BÉBÉ PEUGEOT

A Bébé Peugeot was the tiniest car I ever saw. It was made in France, and only an occasional specimen that had crossed the Alps could be seen in Italy. It burned little more gas than a motorcycle and made the same amount of noise. Because it had no differential and its wheels were obliged to run at the same speed on curves, it moved like a power-propelled baby carriage, jumping and swerving at every turn. The particular Bébé Peugeot of which I am going to talk was a two-seat convertible the color of bright egg-yolk, with a leaky oilcloth roof and a rumble seat in the back. As it sped around at a top velocity of twenty miles per hour, it was always followed by a dense cloud of black smoke from the open exhaust.

I heard of this car some time before I had the pleasure of making its formal acquaintance. It was late September, 1927, and I was staying at my uncle and aunt's villa near Florence. The villa was part farmhouse and part country dwelling. Through a wide-arched entrance the oxcarts reached the wine press and the stone mill for making oil. The space under the roof was used for granaries. The first and second floors were the living quarters, the farmer's in the back, the owners' and their guests' in the front. My entire family, parents, sisters, and brother, were of those guests who went back year after year, and I can't remember a single fall before the last world war when I did not spend a month or two at the villa.

The house, over two hundred years old, was almost ancient even by Italian standards. It stood on the flank of a hill, surrounded by terraced gardens, among thick clumps of shady trees. It looked down on the valley of the Arno River and its neatly cultivated

49

flanks, silvery green with olive trees, lined with rows of intertwining vines, dotted with dark cypresses and colorful peasant houses. Across the valley the mountains of Vallombrosa were close in the evening, distant and uncertain in the morning mist.

It was a perfect setting for study and dream. Early in the morning I always had the graveled terrace under the mulberry trees all to myself. I used to pull a wicker chair in front of the rustic table made of an old grindstone and spread my books on it. I was preparing for a dull exam in organic chemistry. "Methane, propane, butane. . . . Oh! To be on top of that mountain, hand in hand. . . . Methyl, propyl, butyl. . . ."

By noon all the other guests had gathered under the mulberry trees, and I kept my eyes on my books and my ears on the chattering around me.

My elderly aunt always had some crocheting along and would add slow stitch after slow stitch to a child's sweater with yarn spun from fleece of the farm's sheep, a static sweater that never seemed to grow and was meant for charity, since my aunt had no children of her own.

My mother, sitting close to my aunt, was forever darning: four children make eight feet, but we could easily have been centipedes, judging from the pile of hose in my mother's basket.

My sister Anna would be painting. She painted all day long: a heap of yellow corn in the granary; the wall that held the terrace and between whose stones the plants of capers grew; the church above the villa, with its procession of dark cypresses pointing to the sky; a peasant child with tow hair and bashful countenance; or my sister Paola, who had gipsy eyes and was olive-skinned.

Other women guests might be sewing or embroidering at a relaxed pace, and my young brother Sandro, if in a genial mood, might pass around some muscat grapes or a few of those sweet, sun-ripened figs that a drop of golden syrup oozing from the bottom makes irresistible.

The older men, my uncle and my father, talked of political issues, the increasing price of the English pound, the abolition of popular elections in Italy, the lack of freedom in our press.

Everybody would now and then turn their eyes on the dusty road that led from the open gate of the garden to the village. On that road and at that time of day, both in rain and in shine, an old woman in faded clothes, carrying a straw bag under her bony arm, would come in sight. She was the mailwoman from the post office three miles away, and more often than not I was the first one to see her in those last days of September, 1927, because, unaware of it myself, I was constantly hoping to receive news of a certain young man of dark complexion whose name was Enrico Fermi.

I had seen him last in August when we had hiked together in Cortina d'Ampezzo. Afterward I read of him in the papers. At the international meeting in Como, where all the greatest stars in physics had gathered, Fermi had talked on some abstruse theory of quantum mechanics, of which I understood nothing. Where he had gone from there, what he was doing at present, I did not know and would have liked to hear.

One day, at long last, the mailwoman brought me news.

"Who's writing you?" My sister Anna had put down her paint-brush to see the mail but had been disappointed to receive no letters.

"Cornelia," I said briefly and plunged into my reading. I hardly heard her say with scorn:

"Always your logarithms!"

A few minutes later she spoke to me again:

"And what does Cornelia tell you?"

"That Fermi has bought a small yellow car and Rasetti a hazel-nut one," I said in one breath, as if I wished to delete the very idea of the yellow car.

"They always do things together, those two, don't they? But why do you make a long face as if something had gone wrong? You should be glad: they'll take you riding in their cars!"

Anna and I, only one year apart in age, had grown up together; she knew my moods better than I did myself.

Not long before, Fermi, the lover of simplicity, of thrift pushed to the extreme, had told his friends that he felt like doing some-thing out of the ordinary, something definitely extravagant: either to buy a car or to take a wife. So he had made up his mind.

"Nothing has gone wrong. On the contrary I am much pleased," I answered my sister. I was sincere. I was determined, then, to be-come a career woman, never to marry. Besides, I knew Fermi's requirements for a wife and was aware that I did not fill them.

When he had once described his ideal wife, following his constant need for definitions, his tone, assured and resolute, had left little doubt that he spoke in earnestness. He wanted a tall, strong girl of the athletic type, and blonde if possible; she must come from sturdy country stock, be nonreligious, and her four grandparents must be alive. It was all in accordance with his ideas on eugenics, his love for sports, and the agnostic attitude that he claimed to have toward the metaphysical and the unknown.

I was neither tall nor blonde nor particularly robust. My only athletic activity was to fall on skis; I had just come out of one of those religious crises through which most girls cannot avoid passing, at least in Catholic Italy; my forefathers, as far back as could be traced, were city-bred and white-collar workers; my father's mother, the last survivor of my four grandparents, had just died in her hundredth year. So far as I was concerned, Fermi was more useful as a driver than as a possible husband.

Fermi, however, was to indulge in greater extravagance than planned: he had bought a car and within a few months he took a wife as well. Thus I came to share the ownership of the Bébé Peugeot.

This happened in due time. Meanwhile, I went back to my chemistry and my dreams. In October I returned to Rome, where I was introduced to the Bébé Peugeot. On Sundays the whole group of logarithms were packed into Fermi's and Rasetti's cars. Often I was asked to climb up the back part of the Peugeot and down into the rumble seat, no easy achievement, what with flying skirts and chivalrous young men proffering their help. The rumble was pleasant in good weather, but when it rained and Fermi pulled the roof over the front seat, leaving us in the back isolated and in the open, the rumble became a humiliating wet hole.

We spent delightful hours scouting the country around Rome, but also much weary time standing still on the side of a road while Fermi and Rasetti huddled their heads together under the hood of one of the two cars, fumbled with gadgets and ignition, and heatedly argued whether the trouble lay in the magneto or in the carburetor.

Its conspicuous color and its foreign make in a country with a booming automobile industry turned the Peugeot into the best-

known car in Rome. During our early married life, whenever Enrico
and I walked to the Peugeot after an evening show, we always
found a jocose note pinned to the seat by one of our friends.

A deft alternation of the automatic starter and of the hand crank
usually put the Peugeot in motion within a reasonable time. Enrico
kept the crank on the seat by him, for he used it too often to part
from it. Nor did he mind this tool's providing an outlet for his sur-
plus physical energy, until a certain cold winter night when he and
I, in evening clothes, had gone to our garage with the intent of
driving to an official party. The automatic starter, chilled in the
unheated garage, refused to work. Enrico grabbed the crank, bent
down with swift determination to insert it in its socket. Crack! His
tuxedo trousers ripped on that part of his body that always was his
plumpest and that in his present position had become most prominent. Red stripes on white cotton came into view. The new tuxedo
that Enrico had to buy offset the thriftiness of the little car.

Because the Peugeot performed spasmodically and was as jumpy
as a bedbug, we usually drove it only in the city or in the immediate
country. But on the first summer of our married life we undertook
a real journey in it and drove to my aunt's villa near Florence. There
are two hundred miles between Rome and Florence, and people in
their right minds used to plan on spending a night on the road. But,
being young and adventurous, we ignored probable repairs and the
Peugeot's slow pace and decided to leave at five in the morning,
hoping to reach our goal not too late the same evening.

Our trip started in a terrific thunderstorm. No matter how tightly
I shut my eyes, the flare of the lightning would reach my retina and
make me jump. Rain came in on us from the cracks in the roof, from
the poor-fitting celluloid windows, from everywhere. Hail rapped
against the windshield, ominously. I was ready to turn around and
go home. But how could I confess my irrational fears to a newly
acquired and all-trusting husband? I clenched my teeth, put a hand
over my eyes, and bravely went on jumping on my seat at each
stroke of lightning and each thunderbolt. We advanced cautiously,
waded into puddles, skidded on mud. In Viterbo we took time to
have a flat fixed. The storm ceased, the sun broke out, baking hot.

Cheerfully we proceeded through the hilly country. By some per-
versity of its own, the road never missed the top of a hill: up and

down it went, from the summit to the very bottom of the valley. Up and down puffed the Peugeot, stopped to catch its breath, staggered on, gained speed coasting downhill. Perched on the top of the highest peak was the old village of Radicofani, the halfway mark between Rome and Florence. The road saw it from afar and rushed in a straight line at the steepest inclination, as if suddenly realizing that it had already wasted too much time. Roaring, the Peugeot darted off from the foot of the hill, spluttered, covered a few yards, and came to a stop. We were in a cloud of steam. Enrico raised the hood and announced that the transmission belt of the cooling fan had disintegrated.

A thousand feet above us the village sneered. We were already two good hours behind schedule: it was two o'clock and we had had no lunch. The sun was scorching, and I was on the verge of tears. But Enrico is a man of resource. He unfastened the belt of his trousers, tied it around the fan, and grinned proudly at me. That night we slept at the villa.

Fermi and Black Sheep   Fermi and bestiolina Nella

The Bébé Peugeot: A Combination of Ignition Key and
Hand Crank Might Start It

The "Cardinal" (Rasetti), the "Pope" (Fermi), and the "Basilisk" (Segré)

The Old Physics Building in Rome

(6)

EARLY MARRIED YEARS

July 19, 1928, was a hot day in Rome—104° in the shade. It was our wedding day.

By ten in the morning relatives and close friends started gathering at my parents' home to proceed later in a group to the City Hall. My mother had given me the advice a mother would, ending with ". . . and see that your husband stops wearing hazelnut suits. They don't become him." The word "husband" and the idea of imposing my will on Enrico even in as trifling a matter as the color of his clothes had seemed strange and startling. But there had been little time to ponder about my feelings. I had put on a dress made for the occasion, the flounciest I ever saw, and set myself to wait for my bridegroom.

The car that had been dispatched to the Città Giardino for Enrico and his sister came back with Maria alone. He was not ready. Soon everybody had come except him. I grew uneasy. When at last he arrived, he explained: he had started to change and put on the brand-new shirt he had bought to go with his morning coat. The sleeves hung down three inches below his fingertips, because shirts were made in one sleeve length only, and care was taken that they should fit the longest-limbed giant. Enrico was alone in the house. What would he do? Unperturbed as usual, he sat by the sewing machine and put a big fold in both sleeves. It was not his first or his most impressive sewing achievement. The summer before in the Alps he had shown me with just pride his knickerbockers which he had made himself, using an old pair for a pattern.

Having successfully solved the problem of the sleeves, Enrico was ready to tackle that of marriage. We piled into cars and left.

We were going to have what was then called "a wedding in Campidoglio," a civil wedding with no church ceremony, because we belonged to different religions. Like most Italians, Enrico was a Catholic, although he had not been religiously brought up; and I belonged to a family of nonobservant Jews.

The cars reached the foot of the Campidoglio, the historical Capitol Hill that in Roman times had been the citadel. There honking geese had roused warriors from sleep and saved the city from Gallic invasion. The cars drove up the ramp cut into the flank of the Rupe Tarpea from whose top Roman traitors were thrown down to their deaths; they passed the small cave that has always housed a live she-wolf, the emblem of Rome; and higher up they drove by a cage, empty now, where in Mussolini's time there was an eagle, because he had revived the war myths of the Latins and the Roman eagle was the symbol of conquest.

The cars stopped on top of the hill, on the Piazza where the much beloved Marcus Aurelius, the philosopher-emperor, forever rides his bronze horse. Three sides of the Piazza are closed by three monumental palaces. One of them is the City Hall. There a city official with a blue scarf across his shoulders administered the wedding oath to us on that hot nineteenth of July.

As we came out from the palace into the square, our wedding group was immortalized in a picture, in the best of traditions. Unluckily for our fair sex, fashions were unsightly, and all women in the group have their faces concealed under deep hats looking like inverted pots. Enrico and I are in front of the group; he grins at the camera man in self-conscious embarrassment, while I awkwardly hold a bunch of flowers. It is the habit for the bridegroom to bring a bouquet to the bride on their wedding day; but on that occasion as always, both before and afterward, Enrico ignored the very existence of flowers. While we were waiting our turn to be married, well-meaning relatives dispatched one of my cousins to the nearest florist.

In that picture it is hard to tell my mother, aunts, sisters, and Enrico's sister apart under their hats, but they were all there. The happily smiling navy officer was my father. The chubby little man with shiny skin on his skull and well-padded face, with witty, dotlike eyes, was Senator Corbino. He had attended the ceremony as Enrico's best man and was the first person to shock me into the realization that I had become a married woman. As soon as the ceremony was over, he walked up to me. With the most solemn expression his jovial face could assume he bowed to me, kissed my hand, and said: "Congratulations, Mrs. Fermi."

Then we went home.

We started our wedding trip with a daring feat: we flew.

Civilian aviation was still very young, and the first passenger line in Italy had been inaugurated only two years before, in April, 1926. At about that time an Italian aviation expert wrote: "Metal construction [of planes] has had such warm advocates that many planes are entirely of metal . . . wood is again taken in consideration because it makes construction easier and less expensive . . . a metallic type has been successfully experimented with . . . which would replace wooden planes."

In 1928 all Italian lines were serviced by two-engine seaplanes of the Dornier-Wall type, built in Italy. If there were reasons for seaplanes to fly on the Palermo-Genoa line that ran along the Italian western coast, it seems strange, to say the least, that they should also have flown on inland routes, like Turin-Pavia-Trieste. Landing facilities in Turin and Pavia consisted of pontoons built on the Po and Ticino rivers, respectively.

The Ministry of Areonautics kept accurate records of those flights, and thus I learned that I was one of 304 Italian women to have flown in the whole year 1928, while Enrico was one of 1,358 Italian men. Among foreigners on Italian lines the proportion of women was somewhat higher: 89 against 242 men, a fact which shows that Italian women are well-behaved and stay at home. Each seaplane could hold as many as eight passengers, but on the average it carried four passengers per flight. In those records, dry as they are, it is easy to detect the great pride of the Italian government in its aviation. The records stress that the flights were held daily except Sundays and holidays; that, once started, they were almost always completed; that there had been no accidents whatever to passengers or crew during the entire year.

The weather was splendid the day we flew to Genoa from the Rome airport on the Tyrrhenian coast at Fiumicino. No cloud in the sky, no wind, no trace of mist over the sea. Our little seaplane followed the shoreline, with its steady succession of fashionable resorts. It flew so low that we could see the colorful umbrellas on the golden sands and the swarms of bathers in the sea, waving their arms to salute the big bird pregnant with adventure. The flight was smooth but for a few bumps here and there that scared me at first, and for the landing, when we went on jumping and bouncing over the sea as if we were never to stop. I was green in the face when we set foot on land, but pleased at having managed to keep my fears to myself. I thought Enrico should be proud of his wife.

From Genoa we went by train and bus to Champoluc, a village in the western Alps, in a valley that leads to the glaciers between Mount Rose and the Matterhorn. Both of us loved the high mountains, the hikes from valley to valley over windy passes, the suddenly revealed majesty of snowy peaks and the green pastures at their feet.

We took hikes that summer and explored that region of the Alps. But after sunset and on rainy days Enrico yielded to the call of his mind. He is a born teacher and cannot do without teaching, so I was to become his pupil then and there. I was to learn physics, all there is to know about physics.

It is Enrico's contention that a good teacher is always successful, no matter how dumb his pupil. He had told me so upon learning that once, in my high-school days, I had tutored two young boys in geometry and that they had both failed. In my defense I pleaded that they had come to me too late, that in the short time left before exams I could not possibly rescue two boys who always started their demonstrations saying: "That this angle is equal to this side," or similar absurdities.

"Nonsense!" Enrico had replied and had proceeded to relate his own experiences: "I was still a small child, perhaps fourteen, and my pupil a big hunk of a boy, older than I. But what an ass!" Enrico used to spend a couple of hours a day with him, helping him digest his mathematics, and the boy passed his exam. Ergo: I was a bad teacher and Enrico an infallible one.

He should have known only too well that a pupil beyond hope is a rare bird, but a real one, and that I was that bird. In the summer we had spent together in Val Gardena I had proved absolutely refractory to learning music. When taking hikes and when climbing down mountainsides, our friends would sing popular songs, and, although I am completely tone-deaf and can't keep two notes straight, I would not deny myself the pleasure of singing along. In despair Enrico and Cornelia resolved to teach me how to sing. I then had two teachers, one who claimed that no teaching job is impossible, the other who had an excellent musical ear and could sing beautifully. By the very end of the summer, and for a day or two, I was able to sing one verse and that one only. The words of that verse were:

"On the stretcher that they are now carrying, there certainly lies my lover's corpse!"

My teachers could think of nothing more cheerful to shout to the fantastic peaks in front of us. When my lessons ceased, I reverted to my personal interpretation of the music of even that one verse.

Now self-confident Enrico was going to teach me physics. I had visions of a life of co-operation, of daily work with a husband whom I had most decidedly placed on a pedestal. I would keep in the shadow, of course, but with my help his pedestal would be raised for all the world to see. . . . He would be indebted to me, and grateful, and loving. . . . Dreams!

In the wooden-lined bedroom of the little mountain inn, where the dampness of a rainy day and the chill of an evening could swiftly chase the balmy warmth of the Alpine sun, Enrico introduced me to the Maxwell equations. Patiently I learned the mathematical instruments needed to follow each passage. Faithfully I went over Enrico's explanations, trying to keep my eyes from the window and the inviting meadow that I saw through it, until I had digested my lesson and made it material of my own brain. Thus we arrived at the end of the long demonstration: the velocity of light and that of electromagnetic waves were expressed by the same number. "Therefore," Enrico said, "light is nothing else but electromagnetic waves."

"How can you say so?"

"We have just demonstrated it."

"I don't think so. You proved only that through some mathematical abstractions you can obtain two equal numbers. But now you talk of equality of two things. You can't do that. Besides, two equal things need not be the same thing."

I would not be persuaded, and that was the end of my training in physics.

Despite the fact that Enrico's expectations as teacher and beguiled husband were thwarted, still I was given an opportunity to help him in a more satisfying way than by just darning his socks.

When we had become engaged, Enrico, conscious of his responsibilities toward his future family, had investigated various means of supplementing his salary, which was about ninety dollars a month and inadequate to provide anything but the bare necessities of life. A university professor in Italy was obliged to rely on extra income. Its most common sources were a family patrimony, a wife's dowry, and royalties from a book. No patrimony had come down to Enrico from his ancestors, so he was left with the other two possibilities. Like most middle-class girls, I had a sum in my name, which I invested in an apartment at the time of our marriage. Nonetheless, Enrico felt that we needed more, not to lead an extravagant existence, but to acquire a sense of security and to be prepared for emergencies. He resolved to exploit the third possible source of income and to write a textbook for *licei*, the Italian high schools.

"I'll dictate it to you," he said. "You can copy it in your spare time and help me draw sketches for the illustrations."

I agreed readily, and we set to the task at once. We resumed it upon returning from our wedding trip, after we had settled in our apartment. Then my job of secretary turned gradually into that of dumb student. To Enrico all physics was "clear," "evident," or "obvious." Not so to me.

"It is evident," he would dictate, "that in a nonuniformly accelerated motion the ratio of the speed to the time is not constant."

Without raising my eyes from the sheet on which I was scribbling in haste to keep up with dictation, I would state:

"It is not evident."

"It is, to anybody with a thinking mind."

"Not to me."

"Because you refuse to use your brains."

How could we ever settle such arguments?

"Let's call Paola on the telephone," I suggested once, and Enrico agreed.

My sister Paola had just passed the comprehensive examination on the three-year curriculum that all students are required to take at the end of *liceo*. For all her dislike for science, Paola had obtained a decent mark in physics, and Enrico could not expect his book to cater to scientific talent exclusively.

Consulted, Paola was baffled.

From then on, she became the sole judge of our disputes over the understandability of physics exposition, and almost invariably she passed sentence favorable to me.

A two-volume book takes a long time to write in any case, but if the secretary interrupts the dictation with exasperating frequency to contest the clarity of each statement, it takes forever. We worked at Enrico's book for almost two years, taking it along on our summer vacations, first to the mountains and in September to my aunt and uncle's villa. In the bedroom overlooking the wide Arno Valley, the bedroom that was always ours during our stays at the villa, we had only a small table for a desk, just sufficient to lay the manuscript down. But Enrico needed to consult no books, only his memory. On that small table in subsequent years he wrote more difficult books, still with little or no reference material, rationing his work—six pages a day if I was not helping, four if I was. There was time to share the pleasant life at the villa.

Enrico was always willing to join the other guests at noon, under the mulberry trees, and talk politics over the latest news brought by the papers. He was glad for the afternoon interruption at five, when we all gathered for tea around the large oval table of ancient oak. We lingered there in the quiet of the fall day, broken only now and then by the distant shout of a peasant urging his pair of oxen to plow his field or by a mooing cow in her near-by stable. The stout maid, walking mincingly on high heels, would pass trays of sweets from her white-gloved hands.

At night, in the living-room, Enrico would always go sit by the ample leather chair in which my uncle relaxed, a cigar in his mouth, his body at rest, his mind always alert and vivid. He liked to talk to young people and was particularly fond of the bright ones. The newly acquired nephew interested him. To my uncle, Enrico always had something to say, although in general he was shy with adults. Both men liked to exchange the factual information that each had and about which both were eager. The problems of the land, the methods of cultivation, alternated in their conversations with the outlook for modern physics and the cost of laboratory equipment. In the end they almost invariably talked finance. My uncle was on the board of directors of an insurance company and owned stock in many industries; but for all his knowledge of the field he sought the plain, common-sense, rational opinion of his nephew, who was struggling with his first investment problems.

The rest of the time Enrico worked. It was at the villa that in the fall of 1929 he received the first batch of a thousand title-pages for his book, which he was to sign and send back to the publisher, according to Italian habit. This habit was probably due to the wisdom of the proverb: "To be trusting is well; not to be trusting is even better." Most authors, however, in whom laziness was stronger than mistrust, printed their signatures with a rubber stamp. To falsify a signature, be it on a rubber stamp, they thought, was such a serious offense that no publisher would ever commit it.

Enrico had procured his rubber stamp in advance, and at once he could tackle the job of signing his title-pages. The small table in our bedroom was not large enough to hold the piles of pages, and we adjourned to the central hall of the villa and to its spacious table. It was on this table that, until a few years before, every night the old chambermaid had prepared rows upon rows of candleholders with their candlesticks, for the guests to take to their rooms—tall, brass candleholders, squat, enameled candleholders, single, double candleholders. I used to pick a different one every evening, and while I read in bed by candlelight, which was forbidden, I nursed my bad conscience and the fear that the mosquito net on my bed might catch on fire.

On that table, under the dim light of the recently installed electric light, Enrico undertook to sign his title-pages. Proudly he brought down the rubber stamp with firm strokes, from ink-pad to paper sheets, while I cheerfully turned the signed pages face down on the table, counting them aloud. Each meant 3.20 lire or 15 American cents, 20 per cent of the price of each volume. Thus the book, although no masterpiece—it was mediocre prose and complied with unimaginative government programs—still served its purpose of bringing economic returns for many years.

In Rome we had settled into a top-floor apartment in a co-opera-tive building. The apartment was pleasant, full of light and air, and more than adequate for a couple. The ceilings in the six rooms were high, the decorations in good taste.

According to current custom, Enrico had provided the money to furnish it. But he had dodged the actual shopping.

"You go ahead and buy what you want," Enrico had told me. "I don't care what furniture looks like, provided its legs are straight."

Straight legs were in line with his taste for simplicity, not only in furniture, but in architecture, food, and clothing as well. Bows, laces, and flounces disappeared from my wardrobe; mustard, mayon-naise, and pickles from our pantry at the time of our marriage.

"Besides," he had stated, "I don't know how to buy furniture."

The trouble was that I did not know how either. For I had re-ceived what Enrico liked to call "an unreal education." Entirely absorbed in my school and in the extracurricular lessons that were to make an accomplished young lady of me, I had neglected all practical sides of life: our maids did the housework, my mother selected my clothes, my parents took care of school tuition and books. I had acquired no feeling for money or how to spend it.

I let my mother help buy our furniture, which means that I accepted as my own the decisions she made for me. She was a woman of sound judgment, and most of our furniture has stood the years and the trip across the ocean. But she let herself be guided by her own aesthetic principles, and some curved legs found their way into our apartment. They were not *too much* curved, and Enrico did not complain. Besides, in his own domain, his study, chair legs were absolutely straight. His study was a small room with a large table and a small bookcase, in accordance with his needs: much space to scatter writing paper and very little for books. To my surprise, he owned few books; and, of these, only perhaps ten volumes he kept at home. The rest were in his room at the physics building, where he worked all day long, except for the two hours before breakfast, from five-thirty to seven-thirty, when he used his study.

Some mornings I would get up from bed a few minutes before seven-thirty and, still half-asleep, I would walk to the study. Enrico, wrapped in his blue flannel robe, perched on a tall armchair with his slippered feet on its front bar, huddled up and bent over the table, would not hear me. He was much too absorbed by his work. But at exactly seven-thirty something snapped in his head, some brain mechanism set like an alarm clock. Then Enrico would come down to earth and stop working. We had breakfast at eight, and immediately after it he left for the university.

The alarm clock in Enrico's brain worked with extreme precision. Enrico was never late and never early for our dinner at one and for our supper at eight. In the afternoon he interrupted reading his paper or playing a game of tennis promptly by three o'clock and went back to work. Experiments in the laboratory had to reach an extraordinary degree of interest before they could upset Enrico's schedule even slightly. Where his brain mechanism failed was in keeping him fully awake until bedtime at nine-thirty. After supper, at the end of a strenuous day, he would go on yawning and rubbing his eyes, but he would bravely wait for the established time to go to bed. Enrico is a man of method.

Our apartment, like all others in our building, had a separate heating plant, a fact not unusual in Italy, where people are great individualists—"quot homines tot sententiae"—and quarrelsome. But the location of the furnace was quite peculiar. To save precious space, the architect had installed it in the maid's toilet, which was just a small cubicle. Our maid was obliged to sit on the bowl to stoke the furnace. She did not mind this; on the contrary, the toilet being the cosiest corner in the house, she chose to spend much time poking at the fire or merely sitting. Once she had acquired this habit, she kept it through the warmer season. In spring and summer the toilet became the seat of her social life, for with the window open she could chatter across the narrow courtyard with the maid in the opposite apartment, who had elected *her* toilet as *her* place of relaxation.

No space to store coal was provided, and our maid was compelled each day to carry several pailfuls of coal from the basement. This fact and her lack of previous experience as fireman explain why our furnace often went out during our first married winter.

It was an unusually harsh winter, the coldest in Europe within man's remembrance. In Rome there was persistent ice on the streets. Water mains froze and burst, causing water shortages in some sections. Most heating plants proved insufficient to meet the extraordinary weather. Our apartment on the top floor of the building benefited from no heat from above. Unluckily, there was no heat from underneath either, for the apartment below us was temporarily vacant, and its furnace was idle. Moreover, the zone of town where we lived was still under construction, and our building, facing open spaces on three sides, was not protected from the winds. When the *tramontana* blew from the north, our home turned into an icebox.

There were days when, despite our maid's long sessions in the toilet, my frantic but inefficient help, and Enrico's theoretical directives on how to keep a furnace going, the temperature would not rise above 46° in our living-room. There was talk of storm windows. Enrico, who always attacks every practical issue in a rational manner, sat at his table in the study and undertook lengthy calculations to determine the amount of cold air from outdoors that could enter through fissures in the window fittings, and its effect on the temperature inside. The result was disheartening: the effect of drafts was negligible, and consequently storm windows would not help. Only several months later could I obtain Enrico's consent to the purchase of storm windows: he had revised his calculations and realized that he had misplaced a decimal point.

This first blatant error of his should have warned me that infallibility is not of this earth, but Enrico was so sensible in the opinions he propounded, so rational in all his statements, that I was inclined to believe he could never be wrong. This belief of mine had some factual justification: Enrico had an amazing power of thinking before talking, of weighing his words with care, of never stating a fact unless he was more than sure of it.

Confronted with so much equanimity, I gradually developed an overconsciousness of my ignorance, of the worthlessness of my own opinions. That feeling was riveted Sunday after Sunday, when hiking with our friends. Almost every Sunday we would take long walks in the country or on the seashore with some members of our group. Emilio Segré, who had always lived in Rome and had other friends, seldom joined us. But Rasetti and Amaldi came frequently. When both Enrico and Franco found themselves in the company of girls, their favorite pastime was to give them an exam in "general culture." Cornelia was good at laughing all questions off as not meant for her. Maria Fermi, a quiet, earnest girl, a scholar in her literary field, would smile a vague, serious smile and look down tolerantly on the others. They would not bother her. Gina Castelnuovo, Ginestra, after she married Edoardo Amaldi, and I were the predestined victims.

If we were walking along the sea at Ostia, on the hard-packed strip of wet sand, where the caressing waves had dragged remains of marine life, Rasetti might pick up a seashell and place it on his open palm. He would look at it a few seconds with probing, dissecting eyes, then he would ask us:

"What is the name of this shell? How does it live?"

If we did not answer immediately, a stream of hard, swift, precise words would rush from his mouth, like a cascade of gravel down the side of a mountain.

"Fantastic!" he would exclaim, "You don't recognize this most common bivalve mollusk. *Tellina pulchella*. Its shell is asymmetric; its two valves are of different convexity. . . ."

When hiking in the country, Enrico might stop suddenly and bend down, pointing his thin nose toward something on the ground. We could see nothing of interest, only a common anthill.

"How many cerebral cells work at building this mound? Would you say that ant brains yield more or less work than human brains per unit of cerebral matter?" Enrico would pull out of his pocket the small slide rule that never left him. "Let's see . . . in a cubic centimeter of neurons. . . ." In a short while he would raise his triumphant eyes on us. "I have figured the answers. And you?"

At other times the "general culture" exams dealt with geography.

"I want to be kind this time," Rasetti might say, "I'll ask you a really easy question. What is the capital of Afghanistan? . . . Fantastic! You don't even know the capital of a country 270,000 square miles in area!"

Fermi showed an astounding ability to figure out an answer to any question posed either by others or by himself. Rasetti had an unlimited store of knowledge. He knew everything: the monastic rules of lamas in Tibet; the time of departure of all European trains; the death dates of all kings of England; the rate of exchange of the Brazilian reis. Omniscience and infallibility! The two drove us crazy.

Ginestra and I planned a counterattack in the end: we would make a study of a specific subject to show off on the next Sunday. As a source of information we would take the *Enciclopedia italiana,* a beautiful and learned work still being compiled at that time. It had been conceived by an industrialist, Giovanni Treccani. In exchange for a golden medal and a seat in the senate, he had spent five million lire to donate the famous fifteenth-century Bible of Borso d'Este to the Fascist government; and he had provided a fund to finance the *Enciclopedia.* Work on the *Enciclopedia* had been a godsend to many scholars, who found in it a means of supplementing their salaries. Enrico had been a paid member on its staff for a time. In 1928 he had been offered a chair of physics in Zurich. Corbino, who wanted to keep him in Rome, managed to get him the position of editor of the section on physics as a compensation for renouncing the Swiss offer.

When Ginestra and I resolved to consult the *Enciclopedia italiana,* only the first few volumes were out, and we were limited in our choice of subject matter to the letter *A.* We decided on Alexandria. What we learned about that ancient city and its life through the centuries silenced Enrico and Franco for a full Sunday —but only for one.

The inferiority complex that I could not fail to develop under those circumstances was unexpectedly cured a few years later. During one of Enrico's summer absences I spent some time in a fashionable resort in the Alps with my sister Paola and her husband Piero Franchetti. Piero, a well-appreciated chemist at a Bemberg factory, had chosen that resort for his vacation in order to join a group of his friends. They were mostly men in industry, and, judging from their subsequent success, pretty good men. There was Giovanni Enriques, an old friend of mine and the son of Professor Enriques the mathematician. Giovanni had a prominent position in one of the best-run Italian industries, the Olivetti typewriter factory. There were members of the Olivetti family; there was the present director of another excellent Italian industry, that of the

Necchi sewing machines. To my great surprise, I found that I could hold my own in conversation with them. We talked as equals about current events, about my travels and theirs, about Enrico's work. And nobody cackled or sneered at me.

Some time later I took one more step in the process of freeing myself from intellectual bondage, when I came to the conclusion that self-assurance is not necessarily a sign of knowledge. It was 1940, and we were settled in the United States. Rasetti, who was teaching at Laval University in Quebec, came to see us, and together we drove to Washington for the spring meeting of the Physical Society.

We had driven some time when Enrico, who never missed a chance to show me his proficiency in Americanism, announced:

"In a short while we'll cross the Mason-Dixon line."

"Mason-Dixon? What's that?" I asked.

"Fantastic! You don't know . . . ," Rasetti started.

"It's the line that divides the North from the South," Enrico explained.

"What sort of line? An imaginary line? A physical line?" I asked.

"It's formed by two rivers, the Mason and the Dixon," Rasetti stated with his usual assurance.

"Rivers! You are entirely wrong!" Enrico sneered. "Mason and Dixon were two American senators, one from the North, the other from the South."

They bet one dollar. It turned out that Charles Mason and Jeremiah Dixon were English astronomers. But Enrico, never a good loser, claimed the dollar.

"Because," he said, "it is conceivable that English astronomers should become American senators; but rivers . . . never."

And so ended the myth of Rasetti's omniscience and of Enrico's infallibility.

(7)

MR. NORTH AND THE ACADEMIES

When Corbino sought the establishment of a chair of theoretical physics at the University of Rome, he encountered violent opposition from one quarter. A professor of advanced physics voiced his resentment of what he called "intrusion" into his own field. What was theoretical physics, he argued, if not a branch of the "advanced"? To profess the need of a new chair, to appoint another man, was tantamount to stating that he was not capable of teaching his own subject. It was a slap in the face. He would oppose Corbino's proposal.

When the chair was established despite his objections and Fermi called to it, the professor of advanced physics openly declared that this was a personal affront. A feud thus originated between him and Corbino that time did not abate, a rift that opened wider as the new group of young physicists invaded the physics building and asserted themselves with the reliance deriving from Corbino's support. The poor man at the opposition was compelled to withdraw into his own domain, a couple of rooms at the northern end of the building. It was from this northern location that the young people received the inspiration for a nickname to be given the advanced physics teacher: Mr. North.

Mr. North, like Corbino, was a native of Sicily. Like Corbino, he was in Messina when, on December 28, 1909, an earthquake destroyed that city. Ninety per cent of the buildings crumbled down, and over 30 per cent of the population was killed. Mr. North had not been so lucky as Corbino, whose wife and relatives had survived with him. The girl whom Mr. North was to marry was killed. So were all his relatives. He was left alone, and a lonely man he remained the rest of his life. Perhaps sympathy and warm human relations might have healed his wounds. But he was inclosed in his shell. His egocentric outlook on the world, the narrowness of his mind, his constant suspicions of other people's motives, kept everyone away and prevented the reciprocal flow of affection that might have helped him.

When I took the first-year course in physics that he was teaching, Mr. North was in his middle years, a soft-spoken, slow-moving man who went about his lectures with calm thoroughness and lack of imagination. He achieved some good results in research, and of these he was unbelievably jealous. His assistant's duties were those of a hired hand, not of a collaborator or even of a helper, and were limited to the opening and closing of electric circuits: "Doctor, open, please." "Please Doctor, close." What was to be closed or opened needed not to be stated, for it seldom varied. But the poor assistant's thinking powers were so dulled by these base activities that in the classroom on Mr. North's bid, "Doctor, open, please," more than once he rushed to open the window rather than the circuit. Mr. North's bulging eyes sent him reproachful glances through thick glasses. But Mr. North never lost his temper and his dignity.

If the assistant ventured to inquire into the nature of the research he was helping with, into the reason for his closing and opening circuits, Mr. North would reply in his polite, soft voice: "You shall read about it in the Reports of the Academy of the Lincei." Ideas must be protected.

Against this lonely professor the young people whom Corbino had gathered around himself directed their criticism and their jokes. And youth is cruel. Suddenly a rumor spread, no one knew from where. Mr. North had the evil eye. Even to mention his name brought bad luck. Hence he should be called Mr. North, exclusively.

It is to be remembered that Mr. North (and Corbino, for that matter) was a Sicilian, and that Sicily is the land of superstitions. Occult powers are on the rampage there.

The rumor found its justification in episodes related by Mr. North himself. From afar he had witnessed one of the major disasters in Italian navigation. He had seen the S.S. "Principessa Mafalda" sink and disappear in the ocean, drowning three hundred

Corbino's "Boys" in 1934: D'Agostino, Segré, Amaldi, Rasetti, Fermi

Members of the Royal Academy of Italy (Fermi, Who Disliked the Feathers and Sword, Is at Far Right)

persons. From very close he had seen a man drop dead unaccountably, on a streetcar platform, at the very moment that he, Mr. North, set foot on it. The rumor was confirmed by events in the very physics building. The best, the most sound, hydrogen tube had exploded for no other reason but the fact that a few minutes earlier Mr. North had cautioned the young men working with it: "Be careful, it might explode."

It is not known whether Mr. North was aware of the reputation he had among the younger people. But something happened in 1928 that many interpreted as revenge, and Fermi was the victim, although he had kept aloof from these petty squabbles, not wanting to waste his time.

Corbino had resolved to propose Fermi for nomination to the Academy of the Lincei, the outstanding Italian scientific and literary academy, at its royal session early in June. That is the only session at which new members are named. However, because a short trip to the United States was to prevent him from attending that meeting, Corbino prepared a letter. And Mr. North readily consented to read it at the royal session.

When Corbino returned from his journey, Fermi was not a member of the Academy of the Lincei. Questioned, Mr. North slapped his forehead regretfully:

"What a poor memory I have!" he exclaimed. "I had forgotten your letter. It is still in my pocket."

Now it was Corbino's turn to score a victory.

As far back as January, 1926, the Italian cabinet had approved the establishment of a new "Royal Academy of Italy." It had been conceived by Benito Mussolini as some kind of superacademy meant to overshadow all existing ones. From its realization the *Duce* hoped for much glory. In line with the centralizing trends of fascism, the new academy was being created "not so much to put in light individual work, as to promote, help with advice and financial support, co-ordinate and channel the intellectual work . . . of the nation."

The first thirty academicians were to be named in March, 1929. According to its statutes, no senators could be named to the academy, and Senator Corbino was excluded a priori. If a physicist were to be included in the first batch of academicians, somebody else would be named. Mr. North's hopes were aroused. His loyalty to the Fascist party was unquestionable and widely recognized. The same could not be said of many physicists of standing. The case for him was strong in his own opinion. He could have had no inkling of what was going on behind the scenes. But in view of the actual nomination, one must conclude that Corbino had his hand in it. And a long-reaching, powerful hand it must have been.

The first thirty names had not yet been made public, but already there was excitement in that part of the physics building closest to Corbino's office. Suddenly a young man rushed with the tidings to the northern end of the building. He had not quite reached Mr. North's office, and already he was shouting:

"The first academicians have been named! There is also a physicist. . . ." At the door Amaldi stopped to catch his breath. Mr. North's bulging eyes popped out further behind his thick lenses. His cheeks were aflame.

". . . Fermi!" the young man exclaimed with candid joy, then he turned away as fast as he had come, but not before he saw Mr. North's face turning purple.

Poor Mr. North nursed his wound a long time. He had few people on his side, only his assistant, the circuit-opener, and a couple of students. But Corbino's group was growing. Corbino racked his brains to find means of financing the "little physicists" who would have liked to come to Rome, and he grabbed as many small positions for them as he could. A young physicist was instructor of mathematics for chemists and scientists, a young physicist was "curator of the tuning fork," the official fork that gave the standard tone to all of Italy. Fermi went along with Corbino in his quest for possible jobs: he accepted a nonpaid post at the Consiglio Nazionale delle Ricerche, which allowed him to hire a paid secretary, and this, of course, was a "little physicist."

These were petty maneuvers, not uncommon in Italy where it is difficult to find money for young people. But in 1931 Corbino succeeded in a greater scheme that gave stability to the Roman school: he diverted a vacant chair of the science faculty to physics and put Rasetti in it as full professor of spectroscopy.

Not even Corbino could go on at his game of creating chairs of physics without encountering opposition. To overcome the faculty's resistance, Corbino had been obliged to make a deal this time. He let the Fascist members of the faculty call to Rome a man whose best qualification as scientist was to be an avowed and vocal Fascist. Yet the deal was advantageous to Corbino, who now had another of his loyal "boys" in the faculty. Besides, by pushing the other young physicists up, Segré in Rasetti's previous place, Amaldi in Segré's, and so on, he made another opening for one more young man.

In his feud with Corbino, however, Mr. North was to have the very last word, years later, after Senator Corbino's death of a violent attack of pneumonia, on January 23, 1937. His sudden and premature death stunned the younger physicists. Before anybody could give much thought to replacing him, before the faculty could gather to name his successor—either Fermi or Rasetti, undoubtedly—Mr. North was appointed head of the physics department and director of the laboratories. The Rector of the university, a staunch Fascist like Mr. North, had the power of bypassing the faculty according to some forgotten rule.

When Enrico was unexpectedly named to the Royal Academy of Italy in 1929, we could not fail to be pleased and excited. He was only twenty-seven years old, and, had Mr. North not aroused Corbino's ire by "forgetting" to present his letter to the Lincei, Enrico would not have received such honor. Not that he cared for honors; he rather shied away from them. All he wanted was to live in peace and to work. But along with membership in the new academy went a handsome salary that would help him to live in peace: it was one and a half times as large as that of the university and could be combined with it. Although Enrico appreciated money for its true worth, the power it has to relieve the strain of making two ends meet, the sense of security that it brings, still he would never seek money, never ask or strive for more of it.

"Money," he used to tell me, "has the tendency of coming of its own will to those who don't look for it. I don't care for money, but it will come to me."

When for several months the salary of the Royal Academy seemed to remain on paper—it started flowing only after the official inauguration of the academy on the following October—I was not the only one to worry. The academy had so far cost Enrico seven thousand lire, or three hundred and fifty dollars, the price of the uniform. This sum was equivalent to three and a half months of his university salary.

Inspired by that of the French Academy, the uniform of the Royal Academy was shiny with heavy silver embroideries, silver stripes on its trousers, and it included a feathered cocked hat, a short sword, and a dark, all-wrapping cloak.

Enrico, who hated to be conspicuous, intensely disliked that attire. He donned it for the first time when the Royal Academy was solemnly inaugurated in Mussolini's presence, on October 28, 1929, the seventh anniversary of the March on Rome. It so happened that a painter was at work in our home that same day, and Enrico felt so embarrassed at the thought that this man might see him in his crazy outfit that he dispatched me to close all doors on the hall and clear the way for him.

The question then arose whether he should call a taxi or drive the little Peugeot, which was not up to the solemnity of the occasion. Faithful by disposition, Enrico concealed feathers and embroideries under his dark cloak, and, more bewildered than proud, he drove off in his yellow shell on wheels.

The inauguration took place in the magnificent Farnesina, the ancient palace frescoed by Raffaello, Peruzzi, and Sodoma, which was restored for the purpose of providing a worthy seat for the Royal Academy.

Another embarrassing feature, besides its uniform, went along with the rank of Academician—the title of "Excellency." Enrico was rather irked than pleased by the notoriety that it produced. Moreover, in his opinion, a title like that is of no use even in a bureaucratic country like Italy, where rank is paramount in any entanglement with red tape.

"If I could say 'I am *my* excellency Fermi' when applying for my birth certificate, I would impress the clerks, and get fast service. But I can't walk up to an office window and say 'I am *his* excellency!' "

I remember a skiing trip when we went to a hotel where we had been once before.

"Are you any relation to His Excellency Fermi?" the manager asked.

"A distant relative," Enrico answered.

"His Excellency comes to this hotel now and then," the manager stated with importance. Enrico achieved his purpose and was "left in peace." He did not have to be introduced to inquisitive hotel guests, and his old snow suit and scuffed ski boots were not stared at.

(8)

A SUMMER IN ANN ARBOR

The process of Americanization should have started for me in the summer of 1930, when Enrico and I spent our first two months in the United States. Enrico had accepted an invitation to lecture on the quantum theory of radiation at the summer symposium for theoretical physics at the University of Michigan in Ann Arbor.

My knowledge of the United States was between fuzzy and nonexistent and if America had been in my history book, I had missed it. While I was in high school, the Fascist school reform caught up with me. One of its innovations consisted in having history and philosophy taught by the same teacher. Before there was time for retraining teachers, part of the pupils were to learn philosophy from a historian, the rest history from a philosopher. I was of the latter, and *my* philosopher knew no history. When called to answer questions, all pupils endowed with some dose of loquacity would talk around the issue; they would all land on the same safe grounds, the causes of the French Revolution. That the American Revolution had been prominent among these was slurred over, perhaps as too factual a cause for the taste of philosophical minds. Accordingly, I knew in a vague fashion that there had been an American War of Independence, but of the Civil War I had never heard until 1930. In this regard I must make a confession.

On the top shelf of my family's bookcase, among old, forgotten volumes, there was one book that had attracted my attention because of its title. It was just a name: *Abraham Lincoln*. I had always been too lazy to climb onto a chair, reach for the book, and read it. So I kept wondering who the man with the strange name could be. In Italy Abraham is an exclusively Jewish name; my family had books on Jewish lore. For these reasons, I assumed that Abraham Lincoln was the Wandering Jew. Only in 1930, when I tried to enlarge, or rather to start, my education in American history, did I learn of the Civil War and of its great leader.

About the English language I had no complexes: I thought I knew it because the process of preparing to be an accomplished young lady had included English lessons. Enrico had only a reading knowledge of English, acquired according to his much-advertised method for learning languages: One starts to read an entertaining book in any entirely unknown foreign language with the help of a dictionary. After the first ten pages one proceeds on his own steam. After reading ten or twelve books, some five thousand pages, one may claim to have learned that language. In order to apply this method to English, Enrico had pulled all the available Jack London books out of a lending library in Rome. The result? Just as much reliance on *his* English as I had on mine. We were both in for a few disappointments.

We landed in New York early in June, and during the few days spent there I experienced all the feelings and sensations of a woman who had hardly ever set foot out of Italy.

I was overwhelmed by the huge city that attempted to grow into the air, toward the skies as well as on the land; by the gray, untidy city that did not know the joy of chattering fountains in the midst of squares or the surprise of an unexpected sight at the turn of a meandering street. In the subway, awed by its efficiency and speed but depressed by its white-tiled, latrine-like stations, I stared into human eyes that belonged to unknown human races; I saw features described in courses of anthropology; I apprehended with a shock the existence of tattoos, not in recitals of exotic storytellers, but on the bare arms of summer riders, in the most civilized city in the world.

Everything was great and big in New York: its skyscrapers and its bridges; the distances people were willing to travel between home and work; the layers of dirt on children's hands and the amounts of wasted paper: paper bags, paper towels and napkins, wrapping paper around cleaned garments, paper that went into making newspapers.

Big and great were the atmospheric manifestations: never before had night come so suddenly, in mid-afternoon; never had I seen so much lightning rip the blackness into fringed shreds, nor had I heard thunder louder than any burst conceived. Thunderstorms that brought the fear of God.

Big and great was the racket in New York. It was Prohibition time, and we heard all the stories that went with it. We were taken to speakeasys and served wine in coffee cups because we were from Italy, the land of songs and wine. And we were kindly but firmly requested to drink our cups of wine to the very bottom, because no alcohol was to be left on the tables, and the truth was of no avail, that we were used to drinking no wine at all.

And big were our struggles with the language. To talk was a difficult problem, to understand and to be understood an insurmountable difficulty. I remember the effort I put into trying to explain to an old shoe-repair man that I wanted my shoes resoled and reheeled; and the sense of futility I experienced on the next day, when he addressed me in Italian, for he had read the label inside my shoes.

After we arrived in Ann Arbor, I found myself running the streets in despair: our sink was stopped up, but I could not say so because from my English novels I had never learned the words "sink" and "stopped up," and unhelpful passers-by feigned not to understand that the "ploombber" I was looking for was just a plumber. We had had to rent a house in Ann Arbor despite our expectations: we had been told, back in Italy, that visiting professors would be housed in a fraternity house. This word suggested only brotherhood, and we had seen in our mind a small pension run by the university in true brotherly love. There was no fraternity, however, for married couples, and I had to keep house.

Meanwhile, Enrico, cool and unworried, was lecturing with his usual self-possession, talking of "even" and "uneven" numbers, of "fiunctions" and "infynite" quantities.

Enrico, however, struck his usual piece of good luck. Two friends volunteered to attend his lectures and at the end of each to give him a list of mispronounced or misused words. Once aware of his errors, Enrico did not repeat them. By the end of the summer he was making only the one or two blatant mistakes that his friends purposely had not corrected, or, they said, his classes would be no fun.

His two friends were George Uhlenbeck and Sam Goudsmit, both from Holland, who had both been Professor Ehrenfest's pupils in Leiden. They were now in the process of becoming Americanized, together with their young Dutch wives. Enrico had met George Uhlenbeck when that lanky six-footer had been in Rome as tutor to the Dutch minister's son. When Uhlenbeck had left Holland for Rome, Professor Ehrenfest had told him:

"There must be in Rome a young theoretical physicist, a certain Fermi. He has published some well-thought-out papers in the *Zeitschrift für Physik* on statistics and quantum mechanics. He must be about your age. Why don't you look him up?"

Uhlenbeck had looked up Fermi and seen him occasionally during the winter of 1924, when the grave duties of a tutor gave him free time. Next fall Enrico had gone to Leiden for three months, and there he had met Ehrenfest and Goudsmit.

The Dutch group was friendly and congenial, and from them Fermi had drawn the reassurance that he had needed and that he had not found previously in Göttingen. Professor Ehrenfest, who brought to his teaching a warm human interest in his pupils and the understanding of a loving father, had told Fermi what he wanted to be told, that he had the stuff of a good physicist.

A few years later Uhlenbeck and Goudsmit had both settled in Ann Arbor. While George and Else Uhlenbeck took life in the United States with cheerfulness and the spirit of adventure, Sam and Jeanne Goudsmit seemed wrapped in a veil of nostalgia that softened their moods. The 1930 summer session was a happy time for them. Professor Randall, the head of the physics department, had accepted their suggestions for people to be asked to the symposium. Professor Ehrenfest had come from Holland, and younger Dutch physicists had also gathered in Ann Arbor.

Drawn as they were toward anything that reminded them of their home, the Goudsmits played host to the Dutch group and to other visitors of that summer, and we often spent an informal evening in their home.

An English conversation was hard for me to follow, and at the Goudsmits' it was even harder. Often, when English became less intelligible than usual, I realized that the group had shifted to Dutch without notice. And when English lost any meaning altogether, they were talking German. I would switch off my attention and keep quiet in my chair, busy with my own thoughts.

On such an occasion I was abruptly awakened from daydreaming by a sudden commotion. Around me all were jumping to their feet from their chairs or from the carpet where they had been sitting. Professor Ehrenfest, stocky, with frizzy thick hair and bushy gray mustache, was hurrying to the door with clumsy motions, like a nice old bear. Sam Goudsmit was already out, opening the car, and tall George Uhlenbeck motioned the ones behind with inviting gestures of his huge hands.

"But what? . . . But why? . . ." I tried to ask.

"Hurry up. Come along. . . ," George answered. Enrico had gone. He was probably in Sam's car, which sped away as I got into the Uhlenbecks'. We rode swiftly through the country. Five miles? Ten miles? I gave up trying to take in the events intelligently. And then we stopped.

There was a fire. An old barn was burning, with glowing tongues of fire and heavy columns of smoke.

The awe of the miracle of fire, inherited through generations of men from primitive ancestors and revived by the siren of a fire engine, had urged the group, pressed them in the wake of motorized firemen, to witness the ever new portent of burning things.

Other symptoms of primeval instincts were evident in Ann Arbor. The inborn trust in the bosom of the earth as the only place where one's possessions are safe from the greed of others had caused the chemistry department to start digging three stores underground. At the lowest level they were going to store their allotment of alcohol for scientific purposes, because in Prohibition times men were known to have sold their souls for a drink of alcohol.

The deep-rooted certainty that man is venal, that everything, including justice, must be paid for in good money, had kept alive, and still keeps alive in our days, an ancient habit partaking of primeval cruelty: the habit of setting a price for the capture of an outlaw or for information about him. Such things have long ceased to exist in Italy. To one coming from a country often depicted as the land of vendettas there was a perverse flavor in this more refined, this pondered-upon vendetta, this revenge ordered from a desk in an office and carried out in cold blood.

There seemed to be total incomprehension of some instinctive human feelings in the Americans' insistence on separation of the sexes, asking husbands to stag dinner parties, leaving poor young wives to mope at home; or planning women's lunches, where the same poor wives were to find their way among strangers speaking an idiom strange in words and meanings, without the much-needed support of those pillars of strength, their husbands.

Three questions were asked us again and again that summer in Ann Arbor. "How long have you been married?" "What do you think of Signor Mussolini?" and "How do you like America?"

The first question was not controversial and easily answered: "Two years." The second, asked with a benevolent smile, was indicative of the great interest that fascism had aroused in the United States. Those were good times for fascism, which was looked upon with tolerance, often with sympathy, both inside Italy and abroad. It is true that all freedoms were progressively being abolished; that all powers were slowly concentrating in one man. But, because of their very slowness, these processes were little felt and not much opposed. The largest part of the Italian population was conditioned by the daily Fascist press hammering with great fanfare on the achievements of the Fascist regime.

Thus the man with the forward-jutting chin, who early in the morning rode handsome horses in the Villa Borghese and dispensed parsimonious smiles to admiring girls, who had a histrionic power of arousing the masses with his crude, violent speeches, was still popular in 1930. Only a year before, on February 11, 1929, his prestige had received a sharp boost, when he concluded the *Concordato*, the reconciliation with the pope.

The rift between the Italian state and the Catholic church was then fifty-nine years old. It had started on September 20, 1870, when the papal troops had capitulated and surrendered Rome to the victorious army of Victor Emanuel II, fighting for the unity of Italy. Then Pius IX had withdrawn inside his Vatican palace, where he and his successors had been voluntary prisoners ever since. From there they had waged their spiritual war against the kingdom of Italy.

Mussolini had been able to settle a dispute that had seemed insoluble: he had achieved for the first time the spiritual unity of Italy. It was a great feat, hailed by the millions of Catholics the world over.

I do not remember how we answered the question "What do you think of Signor Mussolini?" We probably smiled back at our acquaintances, with their same benevolent smile, and did little to reverse the good opinion that most Americans showed for fascism.

Then there was the third question: "How do you like America?" Of course we liked it, and stated so with warmth and sincerity. Who would not like a country where everybody is kind, hospitable, and helpful to foreigners, where nobody laughs at them for their awkwardness and their mistakes? However, when I now re-examine my feelings at that time, I realize that I had not grasped the significance of America and of its great institutions. Using the wrong yardstick, I had tried to measure the Americans by traits that would compare with the Europeans' and thus failed to see their basic qualities. I mistook their spontaneity, their lack of constraint and inhibitions, for immaturity. I did not realize that the European refinement, which I missed in the American ways of life, may well have been an indication of decadence; that in the United States that refinement of the few had been replaced by higher living standards of the masses; that by accepting the principle that all men are born equal and have the same right to happiness, the Americans had renounced many of the old country's privileges.

Enrico went back to the United States in the summers of 1933, 1935, 1936, and 1937. Busy with my children, I never followed him again until we came for good. At each trip he came to like the United States better, to appreciate and understand the American people more deeply. At the same time he had the opportunity of looking at Italy and at fascism from the outside, of gaining a perspective that one inside entirely lacked.

Upon his return from each trip, before he became again so enwrapped in his work that nothing else mattered, he would talk of moving to the United States, of escaping from a dictatorship into a democracy. I was always against it. I opposed a change that I did not believe was for the best. The summer I spent in Ann Arbor in 1930 had not started the process of Americanization for me.

(9)

WORK

In January, 1934, the French physicists Frédéric Joliot and his wife Irene Curie announced that they had discovered artificial radioactivity. They had been bombarding aluminum with fast alpha particles. They noticed that the product of disintegration was not stable but that within a few minutes it emitted small particles (positrons), thus behaving like a radioactive substance. Not only aluminum but also a limited number of light-weight elements transformed into radioactive substances under alpha bombardment. On heavier elements alpha particles had no effect.

Alpha particles are positively charged helium nuclei. Their efficiency as nuclear projectiles is limited by their positive charge, which acts as a double obstacle: on the one hand, the attraction exerted on them by the negatively charged electrons surrounding all nuclei slows them down so rapidly that soon they are altogether stopped. Their chances of encountering a nucleus on the short path they travel are exceedingly low. On the other hand, if an alpha particle manages to come in contact with a nucleus, the impact of the collision is greatly reduced because both target and projectile are positively charged and they repel each other with a force that is enormous when the distance between them becomes very small. The number of electrons and the positive charge of the nucleus are larger in heavier elements, a fact explaining why bombardment of heavy elements with alpha particles produces no results.

After Enrico learned of Joliot and Curie's discovery, he decided he would try to produce artificial radioactivity with neutrons. Having no electric charge, neutrons are neither attracted by electrons nor repelled by nuclei; their path inside matter is much longer than that of alpha particles; their speed and energy remain higher; their chances of hitting a nucleus with full impact are much greater. Against these unquestionable advantages, neutrons present a decidedly strong drawback. Unlike alpha particles, they are not emitted spontaneously by radioactive substances, but they are produced by bombarding certain elements with alpha particles, a process yielding approximately one neutron for every hundred thousand alpha particles. This very low yield made the use of neutrons appear questionable.

Only actual experiment could tell whether or not neutrons were good nuclear projectiles, and Enrico resolved to turn into an experimental physicist. He felt the need for a vacation from theoretical work, having just completed an abstruse theory on the emission of beta rays from nuclei in natural radioactive processes. This theory was soon considered one of his major works, but at the moment it was causing annoyance and disappointment. The scientific magazine *Nature,* to which Enrico had sent his paper, had turned it down with the statement that it was not quite suited to that magazine. Enrico's "Tentative Theory of Beta Rays" was consequently published in Italian, in the *Ricerca Scientifica* and in the *Nuovo Cimento,* and soon afterward in German, in the *Zeitschrift für Physik,* but not in English. A switch from theory to experiment was in order and would provide a pleasant diversion for a short while. Enrico could not foresee that he was to start a series of investigations that would keep him busy for years and eventually lead the German scientists Hahn and Strassman to discover uranium fission.

Enrico had done experimental work before, but neither he nor anyone else in Rome had ever attempted nuclear transformations. Rasetti was the chief experimentalist in the Roman group, and, because his main interest was spectroscopy, he had trained the others in that field. Fermi had collaborated with Rasetti and had done both theoretical and experimental research in spectroscopy.

Spectroscopy is of no help to one who wants to bombard atoms with neutrons. Enrico had to learn new techniques, to procure a source of neutrons and a device to detect the products of disintegration. Such a device is a Geiger counter, one of the most common items in the equipment of modern laboratories. In 1934 Geiger counters were still a novelty, known to a few, not available for purchase. The only way to obtain Geiger counters was to make them, and Enrico did not know how.

Rasetti could have helped a great deal, for Rasetti was excellent at building equipment. In making and fitting parts, in handling glass, in connecting the finest wires, he brought that same precision, that same delicate touch of his deft fingers with which he prepared the most fragile wings of Coleoptera for his collection; with which in later years he separated fossil trilobites from the inclosing rocks, to gather a collection second only to that of the Smithsonian Institution. But restless Rasetti was traveling in pursuit of an elusive phantom, the hope of finding something, somewhere, that would give him satisfaction and the sense of fulfilment. Rasetti was in Morocco for a long vacation.

Enrico tackled the task of making counters by himself, and he had them ready in a reasonable time. He still needed a source of neutrons. It was bestowed on him through the intervention of the "Divine Providence." This was embodied in Professor Giulio Cesare Trabacchi, director of the physics laboratory of the Sanità Publica ("bureau of public health") which was housed at that time in the physics building of the university. The Sanità was somewhat richer than the university, and Professor Trabacchi had a more comfortable budget than Professor Corbino; moreover, being an extremely orderly man, he always had everything on hand. Whenever one of our physicists was in need of something, be it a screwdriver or a source of neutrons, Trabacchi was always able to provide it. The grateful young people called him nothing else but "Divine Providence."

In the basement of the physics building Trabacchi kept a gram of radium belonging to the Sanita and an apparatus to extract radon from radium. Radon is a gaseous substance which is formed in the natural disintegration of radium. In its turn, it disintegrates spontaneously, emitting alpha particles. If radon is mixed with beryllium powder, the alpha particles strike the beryllium and cause it to emit neutrons. In making his radon available to Enrico, the Divine Providence gave him a good source of neutrons.

Now Enrico was ready for the first experiments. Being a man of method, he did not start by bombarding substances at random, but proceeded in order, starting from the lightest element, hydrogen, and following the periodic table of elements. Hydrogen gave no results: when he bombarded water with neutrons, nothing happened. He tried lithium next, but again without luck. He went on to beryllium, then to boron, to carbon, to nitrogen. None were activated. Enrico wavered, discouraged, and was on the point of giving up his researches, but his stubbornness made him refuse to yield. He would try one more element. That oxygen would not become radioactive he knew already, for his first bombardment had been on water. So he irradiated fluorine. Hurrah! He was rewarded. Fluorine was strongly activated, and so were other elements that came after fluorine in the periodic table.

This field of investigation appeared so fruitful that Enrico not only enlisted the help of Emilio Segré and of Edoardo Amaldi but felt justified in sending a cable to Rasetti in Morocco, to inform him of the experiments and to advise him to come back at once. A short while later a chemist, Oscar D'Agostino, joined the group, and systematic investigation was carried on at a fast pace.

Enrico would have liked to examine the results of neutron bombardment on all the 92 elements existing on the earth, but several of them are rare and not readily available. Enrico placed Emilio Segré in charge of procurement and urged him to do the best he could.

Edoardo Amaldi, the son of a university professor, and Fermi had learned from their families how to save money, not how to spend it. Enrico disliked shopping intensely. When he could not possibly avoid a purchase, he would go into a store, ask to see all available items in the line he needed, and then buy the least expensive on principle.

Emilio Segré was the son of a businessman who owned a paper-mill near Rome. From early childhood Emilio had heard talk of money, of sales, of purchases, and of investments. He had soon developed not only an interest but also a flair for low and high finance. Consequently, in the team at the physics building he was the best qualified for procurement.

To get all elements he could possibly obtain, Emilio set forth with a shopping bag and a shopping list scribbled by Enrico on a scrap of paper—the periodic table would not do, because many elements would be more conveniently handled in a compound rather than pure. Emilio went to see Mr. Troccoli, the main supplier of chemicals in Rome. Mr. Troccoli had been born in the not too advanced country south of Rome, but, despite this fact, he could converse in Latin with Emilio, having been raised by priests.

Mr. Troccoli proved helpful to an unexpected degree, and Emilio could check off his list and drop into his shopping bag almost all substances that would be available from the best suppliers in present days. Moreover, Mr. Troccoli showed unusual lack of interest in pecuniary gain. When, in going down his list, Emilio reached cesium and rubidium (two soft, silvery metals seldom used in chemistry), Mr. Troccoli got them down from the highest and dustiest shelf, saying:

"You can have these free. They have been in my store for the last fifteen years and nobody has ever asked for them. *Rubidium caesiumque tibi donabo gratis et amore dei.*"

A much pleased Emilio walked back to the physics building, bending under the weight of his full shopping bag, up the graveled alley from Via Panisperna.

The extraction of gaseous radon from radium is a delicate operation, as I had an opportunity to see with my own eyes. Ginestra Amaldi and I had gone to meet our husbands at the physics building and had found it apparently deserted. Nobody was in sight. At last we ran into Rasetti, who was wandering all by himself on the second floor, where offices and laboratories were located. He told us that the others were extracting radon in the basement, and he volunteered to lead us there.

Enrico, Emilio, and Edoardo in grimy laboratory coats were busying themselves around a complicated apparatus made of vertical glass pipes several feet tall. They paid no attention to us but went on with their work.

"In the safe that you can see behind those pipes the Divine Providence keeps his gram of radium," Franco explained to us. "It is worth about six hundred seventy thousand lire" (about thirty-four thousand dollars, at that time).

"Radium disintegrates spontaneously," Franco went on, "and emits radon. While the radium remains inside the safe, the radon, which is a gas, is led by glass pipes passing through the wall of the safe into the outer apparatus. There it goes through purifiers and chemicals, and then it is ready to be extracted.

"Enrico and the others have placed some beryllium powder in that tiny glass tube, hardly over half an inch long, that the Pope holds in his hand, and they will try to fill it with radon." At this point Rasetti raised his voice to add: "But it will break."

Rasetti had' not offered to help the others and was standing in the doorway, his hands in his pockets, an expression of mild mockery on his tanned face. He had returned from Morocco a few days before.

Enrico looked at him, pressing his thin lips tightly, annoyed.

"The Cardinal is in one of his restive moods," Edoardo stated with tolerance. "He will get over it. He will overcome his reluctance to co-operate and will soon work with us like a good boy. Or else I'll spank him, as he knows by experience." Edoardo was no longer the cherubic-looking student, but a rosy-cheeked, well-balanced young married man, and the only person able to cope with Rasetti's tempers.

Rasetti did not heed him, and proceeded with his explanation to Ginestra and me:

"Emilio is now immersing the bottom of the small glass tube in a container filled with liquid air, so that radon shall condense as it flows inside the tube. Otherwise it would escape, as it is a gas at room temperature. They'll break the tube anyway."

Enrico had gone to light a gas flame at the other end of the room and had his back to us. I would have liked to see his face. Emilio's lower lip fell forward, a sure sign that he was displeased.

Amaldi opened the stopcock and let the radon out. His chestnut head and Emilio's dark one huddled together as they watched the drops that formed in the small tube.

"It's ready," Emilio said and rushed the tube to Enrico.

"They want to seal it, but it will break anyhow," Rasetti said with testiness.

"I'll spank you," Edoardo admonished without turning his eyes on him.

But Emilio, the Basilisk, sent out incendiary glances.

Then, in the brief silence that followed, "pop" went the little tube.

It is to his friends' credit that Franco did not come to share Mr. North's reputation of having the evil eye.

The activity of radon decayed in a few days, and fresh radon was extracted once a week. The small glass tubes that could be sealed without breaking were brought to the second floor and used as neutron sources to irradiate all the elements that Emilio had procured. Irradiated substances were tested for radioactivity with Geiger counters. The radiation emitted by the neutron source would have disturbed the measurements had it reached the counters. Therefore, the room where substances were irradiated and the room with the counters were at the two ends of a long corridor.

Sometimes the radioactivity produced in an element was of short duration, and after less than a minute it could no longer be detected. Then haste was essential, and the time to cover the length of the corridor had to be reduced by swift running. Amaldi and Fermi prided themselves on being the fastest runners, and theirs was the task of speeding short-lived substances from one end of the corridor to the other. They always raced, and Enrico claims that he could run faster than Edoardo. But he is not a good loser.

While these experiments were going on, a respectable-looking Spanish scientist in black suit and white shirt came one day to the physics building and asked to see "Sua Eccellenza Fermi." Emilio Segré, who happened to be in the hall on the first floor, told him absent-mindedly: "The Pope is upstairs," and upon noticing the other's puzzled expression he added: "I mean Fermi, of course."

As the visitor reached the second floor, a rosy-cheeked youth and a short-legged man, both in dirty gray coats, tore madly by him, holding strange objects in their hands. Bewildered, the visitor wandered around a while, found nobody in sight, and came back to the hall. Again the two madmen tore by him. At last he found Gian Carlo Wick, a soft-spoken, refined young man who was making a name for himself in theoretical physics. Wick was in little sympathy with bustle and fuss.

"I am looking for Sua Eccellenza Fermi," the Spanish man said; "could you show me his office?"

When the visitor and Wick stepped out into the hall, the two strange men were having their third race.

"Enrico," Wick called as loudly as his good manners permitted him. "This gentleman is here to talk to you."

"Come along," Enrico shouted and disappeared.

The interview took place in front of a counter—as all Enrico's interviews did, both with students and with visitors—between readings, while he jotted down figures on bits of paper. But the Spanish visitor could not conceal the depth of his thwarted expectations.

When, after having activated a substance, the physicists wished to determine what radioactive element they had produced, they were faced with the task of separating the active portion from the inactive bulk. There was no hope of applying common chemical methods, because the amount of radioactive elements produced was so exceedingly small that the most accurate chemical tests could not possibly detect it. A fortunate circumstance made the separation feasible: when radioactive atoms of an element are in a solution containing the same element in nonactive form, they will follow it in a chemical separation.

When the group had bombarded iron with neutrons, for instance, they had found that some of it had become radioactive. They thought it probable that the radioactive element that had been produced from iron was no longer iron, but one of the elements close to it in the periodic table. Accordingly, they dissolved the activated iron in nitric acid and added to the solution small amounts of chromium, manganese, and cobalt. Usual methods of chemical separation were then followed, and the separated elements were tested with Geiger counters. The activity accompanied manganese, and the physicists could then assume that when iron is bombarded with neutrons, it transforms into manganese.

When, in the course of their researches, they came to bombard with neutrons the last element of the periodic table, uranium, whose atomic number is 92, they found that it became active, that more than one element was produced, and that at least one of the radioactive products was none of the existing elements close to uranium. Theoretical considerations and chemical analysis seemed to indicate that a new element of atomic number 93, an element which does not exist on the earth because it is not stable, was among the disintegration products of uranium.

They sent their first report to *Ricerca Scientifica* in May, 1934, not to claim the discovery of a new element but rather to relate what indications they had found that such an element might have been produced.

On June 4, Senator Corbino gave a speech at the royal session of the Academy of the Lincei, in the presence of His Majesty the king. After reviewing the status of modern physics, he came to describe in detail the experiments carried out in Rome. Always a brilliant orator, he spoke on this occasion with enthusiasm and passion, for he was genuinely proud of "his boys," as he called the young physicists. He held the attention of the audience. When he arrived at the work with uranium and the possible creation of element 93, he admitted that Fermi's circumspection in pursuing further investigations before announcing the discovery as final was justified. He went on to state that "from the progress of these investigations, which I have followed day by day, I feel I can conclude that production of this element has already been definitely ascertained."

In the days immediately following Corbino's speech the Fascist press made much ado of the "Fascist victories in the field of culture," of the "immense contribution of Italian scientists" to physics, which "proves once more how in the Fascist atmosphere Italy has resumed her ancient role of teacher and vanguard in all fields."

It is quite understandable that the production of a new element should have appealed to the newspapermen more than any other part of Corbino's spech. One second-rate paper went as far as to state that Fermi had presented a small bottle of 93 to the queen of Italy.

Enrico was disturbed. He did not like publicity. He disapproved of Corbino's assurance about the discovery of element 93. He felt that, in any case, the announcement to the public had been untimely, that it should have followed, not preceded, a full scientific report; he feared he would be accused of rashness and levity by physicists in other countries.

I tried to reassure him, to point out that the statement about element 93 had come from Corbino, not from him, and therefore nobody could possibly blame him. But Enrico was right, as always. The press abroad made sensational reports of Corbino's speech.

The *New York Times* published a two-column article under a two-line headline: "Italian Produces 93rd Element by Bombarding Uranium." An Italian paper published a communication from London: "The Roman news of the artificial production of a new element . . . aroused enormous interest in scientific circles. . . . Scientists here do not venture conclusions before receiving further details of the Academician Fermi's work, and declare themselves not to be willing yet to subscribe to the hypothesis presented at the Academy of Lincei by Senator Corbino."

This short item upset Enrico greatly. He read it one evening, and in the night he woke me up, an act absolutely alien to his nature and to his principles. He told me with gravest concern, in a tone that was close to tears, that his reputation was at stake.

"Isn't there anything you can do about it?" I asked, "Couldn't you publish an explanation, somehow?"

"I must consult with Corbino in the morning," Enrico said. And it was typical of him. Corbino had been the cause, if unwittingly, of his predicament, but this fact had not affected Enrico's judgment. He held Corbino to be a man of acute critical sense, of unquestionable scientific integrity. Corbino was the Maestro, always ready to give sound advice, based on profound knowledge of human nature, on keen discernment, on personal experience. When in need of help, Enrico had always turned to Corbino, and he was going to turn to him this time also.

On the next morning he and Corbino prepared a note for the press that read in part:

"The public is giving an incorrect interpretation . . . to Senator Corbino's speech. . . . It has been ascertained in my researches that . . . many elements bombarded with neutrons change into different elements having radioactive properties. . . . Because uranium is the last of the elements in the atomic series, it appears possible that the element produced should be the following, namely, 93. . . . As is clearly stated in Senator Corbino's speech and in the preliminary notes to scientific magazines . . . numerous and delicate tests must still be performed before production of element 93 is actually proved. . . . At any rate the principal purpose of this research is not to produce a new element, but to study the general phenomenon."

The controversy around element 93 went on with various vicissitudes, as physicists of different countries successively confirmed or refuted its discovery.

Corbino's faith in "his boys" never wavered, nor his belief that they were the first men to make new elements. Only one month before his death he said in a speech published in *Nuova Antologia:*

"This discovery was doubted with real levity. . . . But recently the two greatest experts in radioactive chemistry, Lise Meitner and Otto Hahn of Berlin, have fully confirmed Fermi's discovery. Thus the reservation made by the discoverer in 1934 can be entirely lifted."

But the last word on element 93 had not been said.

(10)

SOUTH AMERICAN INTERLUDE

The Roman summer never fails to disrupt research. The persistent heat and the scorching sun cut down the will to work, drive all who can possibly leave to pleasanter climates in the mountains or by the seashore. In 1934 there were other reasons than inclement weather for an interruption. Enrico had accepted an invitation to deliver lectures in Argentina and Brazil, and he could not break his engagement at the last moment to pursue experiments, no matter how absorbing and promising.

It would have been a mistake to forego this trip, which proved most successful from all points of view. Sixteen days on placid seas took us to Buenos Aires, and there we lived the life of the elite for over three weeks. We were housed in the most modern and elegant hotel we had seen up to then. Introduced by the Italian ambassador and by the president of the Instituto Argentino de Cultura Italica, the sponsor of Enrico's visit, we were entertained in the upper spheres of New World success and wealth. Perhaps out of true interest in science or because of a vague nostalgia for the culture of the Old World that they could not forget, many prominent citizens of Buenos Aires seemed eager to shed their kindness on us. They took us for rides along the Rio de la Plata, and up to the Parana; they invited us into their theater boxes for the best shows and musical performances; they entertained us in their sumptuous homes with that proverbial Spanish hospitality, so hard on the guests' digestive system, that makes a hostess place another guest in charge of keeping your plate full at a meal of five courses—we came to look forward to the rare occasions when, free of invitations, we could quietly skip a meal.

A great testimony of interest in science was given to Enrico.

He held his lectures in halls that were crowded and overflowing at the onset and kept overflowing to the very end of his course, despite the fact that he lectured in Italian. Spanish and Italian have much in common; moreover, a good portion of the Buenos Aires population is of Italian descent.

The lecture halls were tightly packed not only in Buenos Aires but wherever else Enrico gave a talk that summer: in small Córdoba, the town of the many churches at the feet of the Andes, where the only Italian was a fencing teacher; in orderly, green-gardened, intellectual Montevideo; in São Paulo, in whose surroundings the intense green of the tropical vegetation sprang from the bright-red soil that gives Brazil its name in an antithesis of colors seldom achieved by painters; and in Rio de Janeiro.

Only in Rio de Janeiro of these South American cities do I remember to have seen conditions indicating that well-being was not then prevalent, not shared by all. The contrast between opulence and indigence, between good and poor health, was then impressive. Even more astounding was the constant apprehension under which many lived of contracting one of the dreadful tropical diseases, leprosy or trachoma, for instance, through a stranger or a beggar who might chance to accost them. Proof of this apprehension, which sometimes reached fear, were the endless tales of how lepers had tried to infect others with their illness in the belief that they would be cured if seven sound people should acquire it; of how leprosy was rampant everywhere, undetectable in its first stages; and of the difficulties encountered in attempting to isolate all contagious cases, due to the fact that many among the aristocracy had the disease and aristocracy could not be forced inside a leprosarium. Evidence of the Brazilians' determination to drive illness out of their country was the very modern and well-equipped Institute for Tropical Diseases.

We left South America happy to end an experience too rich to be stretched any further. Of good things one must not have too much, and, as the French say, one may tire of eating partridge every day.

On boarding the boat that would take us from Rio to Naples, we met the composer Ottorino Respighi and his wife, and we learned that they would be our companions for the trip. Enrico and Respighi knew each other slightly, because they were both members of the Royal Academy of Italy; but I had met neither him nor Mrs. Respighi.

Enrico, who never misses a chance of what he calls "extracting information" from a new acquaintance, kept Respighi under a steady fire of questions about music. Enrico's approach to the realm of sound made Respighi smile with condescension. He displayed the patience used with inquisitive children, for Enrico tried to have the composer reduce music for his sake to a set of mathematical correlations, a sequence of measurable, numerical intervals, a pattern of vibratory waves that ought to be reproducible by a a drawing on paper. In his turn Enrico smiled with tolerance at the older man's description of his working habits, by fits, at the oddest hours, often in the middle of the night, or halfway through a meal, whenever the inspiration surged in him, impelling, urging, irrepressible. And the smile on Enrico's face was accentuated when Respighi told him how his only truly scientific endeavors had been experiments with a divining rod, both to find water and to detect metals that his wife had hidden for him under the living-room carpet.

During the two weeks at sea we were the closest of friends, but afterward we saw them only once or twice in the few years before his premature end. This is what Enrico likes about sea voyages, the easy coming together and parting, the sudden friendship, the intimacy that, in his opinion, it would be tiresome to keep up on land. And this is what I dislike, because in parting I feel that what I received and gave, what I would like to go on receiving and giving, is lost, and only a sense of futility remains. But such is the difference between man and woman, that only woman likes to keep her hold on what she has grasped, like an oyster does on a rock or a shoot of ivy on a fence.

In pleasant company days passed swiftly, and we arrived in Naples toward the end of September. Enrico and I proceeded to my aunt's villa near Florence, where each year we spent part of the fall months and where we had left our three-year-old daughter Nella and her nursemaid. Soon Enrico returned to Rome alone, while I lingered on at the villa. Thus I happened to miss a great deal of excitement.

(11)

AN ACCIDENTAL DISCOVERY

While we were traveling in South America a physics student in Rome had received his degree and had joined the group of researchers. He was Bruno Pontecorvo, the man who vanished behind the Iron Curtain sixteen years later. Bruno, twenty-one years old in 1934, was one of a large clan of Pontecorvos, brothers, sisters, cousins, living in Pisa. When Rasetti was at the university in that city, he had known the family well and had been the friend of some of the older Pontecorvo boys. At that time Bruno was a small child, "the cub," to whom the rest of the clan paid little attention. The child had grown, gone to school, and, after two years at the university in Pisa, had resolved to study physics in Rome.

When he had called on Rasetti to tell him of his intentions, Rasetti had looked him over:

"Do you really mean to say you are the cub? I can't believe it!"

Bruno was uncommonly good looking. Perhaps it was his proportions that made him attractive. No one could wish to broaden his chest and shoulders or to lengthen his arms and legs. Perhaps he had acquired his elegance of carriage on the tennis courts, where he was soon to become a champion. And good manners were natural to him.

"So you want to do physics?" Franco went on teasing Bruno, "Just out of your diapers and you want to do physics. You must hold a high opinion of yourself!" Bruno blushed, as he did at the least provocation, although he spoke with ease and confidence, a contradiction that he never outgrew.

The cub was clearly a bright boy, and there was a tradition of success in the Pontecorvo family. Bruno was accepted as a physics student at the University of Rome, and after he received his degree in the summer of 1934, he was allowed to help the others with their researches in neutron bombardment. This he was doing when Enrico returned to Rome in October.

One morning, Bruno Pontecorvo and Edoardo Amaldi were testing some metals for artificial radioactivity. The metals had been given the shape of hollow cylinders of equal size, inside which the source of neutrons could fit. To irradiate a cylinder, they placed the source inside it, and then set it in a lead box. On that particular morning Amaldi and Pontecorvo were experimenting with silver Pontecorvo was the first to observe that the silver cylinder behaved strangely, that its activity was not always the same, but was different if it had been placed in the middle or in a corner of the lead box.

Baffled, Amaldi and Pontecorvo went to report to Fermi and Rasetti. Franco was inclined to blame the anomalies to statistical error and inaccuracy of measurements. Enrico, who takes an agnostic view of all phenomena, suggested that they try irradiating the silver cylinder outside the lead box and see what happened. More wonders were in store for the next few days. The objects around the cylinder seemed to influence its activity. If the cylinder had been on a wooden table while being irradiated, its activity was greater than if it had been on a piece of metal. By now the whole group's interest had been aroused, and everybody was participating in the work. They placed the neutron source outside the cylinder and interposed objects between them. A plate of lead made the activity increase slightly. Lead is a heavy substance. "Let's try a light one next," Fermi said, "for instance, paraffin." The experiment with paraffin was performed on the morning of October 22.

They took a big block of paraffin, dug a cavity in it, put the neutron source inside the cavity, irradiated the silver cylinder, and brought it to a Geiger counter to measure its activity. The counter clicked madly. The halls of the physics building resounded with loud exclamations: "Fantastic! Incredible! Black magic!" Paraffin increased the artificially induced radioactivity of silver up to one hundred times.

At noon the group parted reluctantly for the usual lunch recess, which generally lasted a good couple of hours. If, as William James maintains, a symptom of neurotic temperament is to do things on the spur of the moment, under pressure of sudden excitement, without postponing action to a more propitious time, Enrico's temperament must be considered neurotic. I am inclined to believe that there must be other explanations of prompt reaction. That lunch recess on October 22 was going to be Enrico's last spent in solitude, for I was going to come back from the country on the next morning. He put his solitude to good use, and by the time he went back to the laboratory he had a theory worked out to explain the strange action of paraffin.

Paraffin contains a great deal of hydrogen. Hydrogen nuclei are protons, particles having the same mass as neutrons. When the source is inclosed in a paraffin block, the neutrons hit the protons in the paraffin before reaching the silver nuclei. In the collision with a proton, a neutron loses part of its energy, in the same manner as a billiard ball is slowed down when it hits a ball of its same size. Before emerging from the paraffin, a neutron will have collided with many protons in succession, and its velocity will be greatly reduced. This *slow* neutron will have a much better chance of being captured by a silver nucleus than a fast one, much as a slow golf ball has a better chance of making a hole than one which zooms fast and may by-pass it.

If Enrico's explanations were correct, any other substance containing a large proportion of hydrogen should have the same effect as paraffin. "Let's try and see what a considerable quantity of water does to the silver activity," Enrico said on that same afternoon.

There was no better place to find a "considerable quantity of water" than the goldfish fountain in Corbino's private garden behind the laboratory. Senator Corbino and his family occupied the third floor of the physics building, a spacious apartment that went along with the position of head of the physics department. The Corbinos had also the use of the back garden, a romantic spot with green foliage and flowers, closed on one side by the wall of the ancient church of San Lorenzo in Panisperna. It was a spot where one might want to take his first love on a spring night and gaze in bliss at the moon through the blooms of the almond tree that overcast the fountain.

There Rasetti had taken his salamanders. He had kept his salamanders in the fountain, and they had laid eggs for him so that he could repeat Speman's experiment: Franco tied a strand of hair around the egg and cut it in two, thus obtaining two embryos out of one egg. On the salamanders he had poured his love and made them the objects of his tender care, until one day they climbed out of the fountain and disappeared.

In that fountain the physicists had sailed certain small toy boats that had suddenly invaded the Italian market. Each little craft bore a tiny candle on its deck. When the candles were lighted, the boats sped and puffed on the water like real motorboats. They were delightful. And the young men, who had never been able to resist the charm of a new toy, had spent much time watching them run in the fountain.

It was natural that, when in need of a considerable amount of water, Fermi and his friends should think of that fountain. On that afternoon of October 22, they rushed their source of neutrons and their silver cylinder to that fountain, and they placed both under water. The goldfish, I am sure, retained their calm and dignity, despite the neutron shower, more than did the crowd outside. The men's excitement was fed on the results of this experiment. It confirmed Fermi's theory. Water also increased the artificial radioactivity of silver by many times.

That same evening they all gathered in Amaldi's home to write their first report, a letter to *Ricerca Scientifica*. Enrico was going to dictate the letter, Emilio to write it, and Ginestra was to type it later. It was all simple and well planned. But the men shouted their suggestions so loudly, they argued so heatedly about what to say and how to say it, they paced the floor in such audible agitation, they left the Amaldi's home in such a state, that the Amaldi's maid timidly inquired whether the guests had all been drunk.

Now they were faced with much more work: again testing a large number of elements; surrounding the neutron source with various thicknesses of appropriate substances; measuring the energies of slowed-down neutrons; perfecting the theory.

One morning, a couple of days after the incursion into his garden, Corbino came to the laboratory; although he did not actively participate in research, he kept informed and often gave good advice. He had followed the younger men's work step by step, and on that morning also he asked to be told what they were at. They were preparing to write a more extensive report on their experiments, they answered. Corbino became incensed.

"What? Do you want to publish more than you have already?" he asked in a swift rush of words, helping the oral expression with brisk gestures, as all Sicilians do. "Are you crazy? Can't you see that your discovery may have industrial applications? You should take a patent before you give out more details on how to make artificial radioactive substances!"

It was a novel idea to them. The six researchers had not thought of themselves as inventors, although they had discussed the possible applications of slow neutrons. The quantities of radioactive elements produced with alpha particles and with regular nonslowed-down neutrons had been so small that no practical use had been foreseen for them. But slow neutrons already permitted production of a hundred times as large quantities. It was conceivable that in the near future artificial radioactive substances would replace the expensive natural ones. The physicists foresaw their use in some medical treatment, in biology as tracers, and as indicators in chemical and industrial processes. The idea of releasing nuclear energy was far from their minds.

As for a patent, they were uncertain. They knew nothing of industrial practices and did not care. They worked in an ivory tower and liked it. Why be bothered? Besides, it was not customary for scientists to take patents on their discoveries. But Corbino insisted: he was a practical man, he had a hand in many industries; age gave him wisdom. . . . The boys were used to following his advice. On October 26, Fermi, Rasetti, Segré, Amaldi, D'Agostino, Pontecorvo, and Trabacchi, the Divine Providence who had furnished the radon for the experiments, jointly applied for a patent for their process to produce artificial radioactivity through slow neutron bombardment.

Work proceeded painstakingly, and a year went by. There were no more spectacular discoveries. By the end of 1935, research and achievement had slowed down, or so believed Emilio Segré, who had enjoyed success and would have gladly tasted more of it. He could see no clear reason for this reduced pace, and, being one who liked to get at the bottom of all questions, he sought Enrico's advice.

"You are the Pope," he said, "and full of wisdom. Can you tell me why we are now accomplishing less than a year ago?"

The Pope showed no hesitation, but replied in oracular sentences:

"Go to the physics library. Pull out the big atlas that is there. Open it. You shall find your explanation."

Emilio complied. Of its own volition the atlas opened at the map of Ethiopia.

The Ethiopian war, which had broken out in October, 1935, after many months of preparation, had preoccupied the physicists like all other thinking Italians. Long before it started, they had taken time to study the maps over and over in search of a justification, or at least an excuse, for a colonial war that had no apparent purpose. In Ethiopia there were no fertile lands, no rich mines, no oil wells, no military bases, no seaports.

From October on, they had followed the unsuccessful campaign on the map, day by day. Together with discontent, with worries about the effects the economic sanctions would have on the already precarious Italian economy, with knowledge that Italy was going from bad to worse, some vague hopes had also arisen. Could serious reverses bring about a political crisis? Perhaps a revolt? A military coup d'état?

How could one give wholehearted attention to research under these circumstances? The mood of carefree co-operation of 1934 was never recaptured entirely. Besides, the group had started to fall apart. In July, 1935, Rasetti went to the United States for a stay of over a year. By the time he came back, Emilio Segré had married and left Rome for Palermo, where he had become professor of physics and director of the physics department. Ettore Majorana, the most brilliant and promising of all the students in Rome, the genius who should have given the greatest contribution to physics, put a dramatic end to his career.

Until 1933 Majorana had been around the laboratory in Rome, working in his peculiar fitful way. Like many great artists, he was seldom satisfied with his work and refrained from publishing anything that was less than perfect. In 1933 he spent some time in Germany, and after that he never went back to work in the physics building.

A tragedy that occurred in the Majorana family may have had

deep effect on Ettore. A baby, a little cousin of Ettore's, was burned to death in his cradle. The baby's nurse was suspected of setting the cradle on fire. One of the baby's uncles was accused of having instigated the nurse. Ettore refused to believe that his uncle could have committed such a depraved, coldblooded crime. Ettore wanted to prove his uncle's innocence, clear him of a suspicion that could not fail to taint the entire Majorana family. He hired lawyers; he took personal charge of all details in the defense. His uncle was acquitted. But the ordeal was a strain on Ettore's sensitive temperament.

After his return from Germany, Ettore became a recluse. He withdrew into his own room in Rome and refused to go out in the streets altogether. A hired woman took care of cleaning his room and getting his meals. Edoardo Amaldi went to see him now and then and tried to make him change his behavior. He used persuasion: he talked and argued in his calm, quiet voice, soothing his school friend of old, asking him to be reasonable, to heed his advice. He alternated persuasion with pressure, he let himself go in the outbursts of rage that occasionally came on him when he was antagonized. Ettore was stubborn. He listened but would not be persuaded. All Edoardo could do for him was to send him a barber who would give him a haircut and a shave in his room.

Meanwhile, other circumstances were ripening that were to be his undoing. A group of young men who had been trained in theoretical physics were very anxious that a *concorso* in their subject should be called soon, so that they could obtain stable university positions. Emilio Segré came to their aid. He was lonely, the only physicist worthy of that name in faraway Palermo; he felt as though his friends had forgotten his very existence. If at least one of the young theoreticians would come and work with him! But what good man would go to Palermo unless there was a substantial compensation for his going?

Segré made a deal with Gian Carlo Wick, who at the moment was in the enviable situation of being able to study and work in Rome, but who, however, had no academic status and would have liked to acquire one. Emilio Segré would see to it that the faculty of science in Palermo should call a *concorso* in theoretical physics. Since Wick was undoubtedly the best of the possible candidates, he would certainly place first and would be called to Palermo. But he would also receive offers from more desirable universities. The price Wick would have to pay Segré for requesting the *concorso* was to pledge himself to stay at least a full year in Palermo. Wick agreed.

The outcome of the *concorso* could be forseen without the least doubt: Wick would place first; Giulio Racah second. The third and last candidate to become qualified to teach theoretical physics in Italian universities would be Giovannino Gentile. He was the son of Giovanni Gentile, the Fascist philosopher and influential politician who had been minister of education in Mussolini's first cabinet and president of the Higher Council for National Education until 1936.

The board of examiners gathered to study the candidates' qualifications. Enrico was on that board.

Then a totally unexpected development upset all predictions. Ettore Majorana announced himself a candidate to the *concorso*. He had consulted nobody, mentioned his decision to no one. The consequences of his move were clear: Majorana would place first, and Giovannino Gentile would be left out of the list of three men qualified for teaching. He would get no stable position until another *concorso* in theoretical physics could be called, probably not for the next few years.

Then something happened that was never known to have happened before. The board of examiners was suspended, temporarily dissolved. When it reconvened a short while later, the minister of education, pressed by philosopher Gentile, had named Ettore Majorana professor of theoretical physics at the University of Naples for special merits and "clear fame," in accordance with an old law revived by fascism. Now the *concorso* could proceed according to the first expectations.

Ettore Majorana went to Naples. Although he had been resolved to teach, he was not able to stand facing his classroom and his students. After a few lectures he left Naples in panic. He boarded a boat for his native Palermo, leaving a suicide note behind. Still he reached Palermo. From there on, all traces of him vanished. One person claimed to have seen him board another boat directed back to Naples. No other passengers saw him, and he did not land in Naples. For a long time his family searched the environs of Palermo and Naples, but he was never found, either alive or dead.

HOW NOT TO RAISE CHILDREN

On January 31, 1931, Nella was born. Although she was a husky baby, Enrico did not dare to take her in his arms or even to touch her. He looked at her from a distance with bewilderment and misgivings and called her *bestiolina*, "little animal."

Babies will get sick occasionally and so did Nella. Whenever Enrico saw her lying down, weak and limp in her cradle, he would become upset.

"These little animals," he would say, "ought to be always well. One cannot bear to see them suffer."

At six months *bestiolina* strongly resembled her father. She even had his expression of one lost in deep thought. I assumed that she would grow to be like him in everything, and I trained her accordingly. During her first summer of life we rented a house in Courmayeur, at the foot of Mont Blanc. Sure that Nella had inherited all of Enrico's traits and would therefore enjoy a climb, I persuaded Enrico to let me take her along on a hike to the Brenva glacier, which comes down on the flank of Mont Blanc to a low level. Her strong peasant nurse did not mind pushing Nella's buggy up the steep path, over rocks and stumps. But I was none too proud of my idea, as I walked with Enrico, behind baby, buggy, and nurse. Every time we passed other hikers I would feel embarrassed and conspicuous, and to avoid their eyes I kept mine on Nella's soft cheeks that were turning ruddy in the pungent air from the glacier.

From her birth on, I kept watching for signs of precocity. But at fourteen months she had to be cheated into walking alone by making her hang on to her own dress.

In one who resembled Enrico I expected early mathematical ability, and as soon as she could talk I looked for it.

"Nella, how many legs does that chair have?"

"Two or three."

While leading her up a flight of stairs, I would prod her:

"Nella, let's count these steps. One . . ."

"One."

"Two . . ."

"Two."

"Three . . ."

"No. This is not three. I told you before that there is no three in these stairs. This is four." Nella has always known her own mind.

She was slightly more interested in economics. Enrico had explained to her the rudiments of budget-making in order to silence her requests for new toys. After that, when she was asked: "What does your father do?" she would answer: "He goes to the laboratory and gets money." "Is that all he does?" "He also divides the money into small piles: one for our food, one for our clothes, one for my toys. One pile he puts in the bank, and there the money grows and grows." As she spoke, she raised her chubby little hand from low down to up above her head, to indicate a fast growth indeed.

She started going to kindergarten a few months before she was five. Her teacher was a young woman who could hardly remember pre-Fascist times and who was driven by a contagious fervor whenever she talked of politics or of religion.

Nella had been exposed to neither. In her relations with us her strong personality had prevailed, and she had led us down to her own level, never attempting to reach the grownups at theirs, never trying to follow their conversations. If we did not talk directly to her, she would turn her attention to a world of fantasy of her own. Politics and religion were not subjects of which we had spoken to her, and of them she knew nothing.

When Catholic religion and fascism were presented to her simultaneously, through the same fanatic devotion, when in her classroom for the first time she saw a Crucifix, the king's and queen's portraits and Mussolini's, all hanging on the same wall, she was impressed and utterly confused.

Upon returning home from school on one of her first days there, she came to see me in my room, and we sat side by side, on the green damask-covered chaise longue, in front of the French window leading to the balcony. The afternoon sun entering the room obliquely through that southern window put a glow of gold in Nella's chestnut hair.

"What did you do in school this morning?" I asked her.

She was wearing the public school uniform: a white cotton dress with a huge sky-blue bow.

"First of all we said our little prayers," she said. Nella's use of words was precise, and her enunciation was as distinct as the sound of drops falling on a stone.

"How many little prayers do you say every morning?"

"One little prayer to the Child Jesus, one little prayer to the king, and one to Mussolini. They will hear them and. . . ."

"The Child Jesus may hear you, but the king and Mussolini are just people like you and Dad and me. They cannot hear. . . ."

"Yes they can," Nella's tone was stern and final.

"If they happened to go by your school," I insisted, "and walked under your windows while you were saying your prayers and if the windows were open, then they could hear you; otherwise they can't."

"I am sure they can always hear me. The teacher would not make me say a prayer to Mussolini if he could not hear it. . . ." Her deep-blue eyes were earnest and eager.

Her confusion of religion and politics was evident at other times. All children in public schools were enrolled *ex officio* in the Fascist youth organization, which was in charge of physical education programs. Five-year-old Nella belonged to the youngest group and was a *figlia della lupa*, a daughter of the historical she-wolf who had nursed Romulus and Remus and fostered the foundation of Rome. Nella was asked to buy a uniform that she was to wear when taking gym.

"I am going to wear my new navy-blue skirt and my white blouse," she said in great expectation the day before her first chance to put it on. "Will you tie my bow for me?"

"Yes, dear."

"I can put on my black cap myself." The fisherman's cap of black silk jersey would look pretty, I knew, on her swinging braids.

"All other little girls will be dressed like me. We'll march: one, two; one, two. Don't you think Mussolini and the king and the Child Jesus will be pleased with us?"

On a Sunday morning she returned from a walk with her nurse in a grumbling mood.

"I wanted to go inside the church where so many people were going, all the Fascists. . . ."

"Do you mean to say the Catholics?"

"Yes, I was mixed up. All the Catholics were going in through those big doors. I wanted to see what's behind those doors. But nursie said no. Why can't I ever go to church? Why don't you and Dad ever go to church?"

At this point I made a mistake. I was overrational. I tried to explain the different beliefs of Christians and Jews, Catholics and Protestants. Nella seemed to follow my words intelligently.

"Do *you* believe that Jesus was God's son?" she asked me at the end of my explanation.

"No, I believe that he was a very good man, who taught people to love each other, but I don't believe that he was God's son."

"What does Dad believe?"

I was not prepared for that question. It is hard to explain to a child the attitude of one who called himself an agnostic, who admitted that with science he might hope to explain almost anything except himself, but who looked at others' spiritual needs with objective rationality.

"Well . . . ," I said, "Dad is a scientist. . . . Like many scientists, he isn't quite sure that God really exists. . . ."

"But is he sure that Mussolini really exists?"

I gave up.

Nella spent much time playing in the *pinetina,* a small grove of umbrella pines at the end of our street, that had been turned into a bit of a public park. Children played on narrow graveled alleys and sickly lawns. The pines were young and cast scanty shadows; their greedy roots sucked up all the moisture of the soil.

Young mothers knitted soft baby things in pink and blue or did some quiet embroidering, while listening to the wisdom of an old invalid, whom a sad-looking daughter wheeled every morning in his chair to the *pinetina.* Apart from mothers, nurses and maids gossiped among themselves on other benches and cast bashful glances at the cock-hatted *carabinieri* across the street, who paced the sidewalk along the Villa Torlonia wall.

Villa Torlonia was Mussolini's residence, and Nella knew it. But the wall inclosing the vast park of that villa was high, taller than any man, and there was no opening on the side facing the *pinetina* or along the street that Nella walked when going to school or to my parents' home. She could see only wall, *carabinieri*, and plainclothesmen. Nella knew of Mussolini's existence behind that wall but had never set eyes on him; he had no more substance for her than the Child Jesus. So I often wondered whether in her mind the Garden of Eden did not look like Villa Torlonia: a high stucco wall the color of sunshine, streaked by rain and age, covered here and there with old, dusty ivy; and, only signs of the marvels and mysteries that could not fail to be inside, the thick foliage of holm oaks, the umbrellas of pine trees, the straight tops of dark cypresses, slender giants keeping a still watch on the park; and, pacing the walks on the outside, cock-hatted *carabinieri* . . . or angels?

Giulio was born on February 16, 1936. He was a strong boy. In the little hospital cradle by my own bed he yelled his lungs into exercise, while white-clad nuns hovered around and vainly tried to pacify him.

Above the baby's crying the voice of newspaper boys often rose from the street, announcing the *Messaggero Rosa*, the pink-paper edition of the *Messaggero*, issued when there was news of extraordinary importance. The *Messaggero Rosa* came out often in those days, to brag of victories in Ethiopia.

"The remains of Ras Mulugheta's army are in flight," the newsboys were shouting under the windows of my hospital room. And "our troops have reached Gaela." And "the Italian colors are on Amba Alagi."

It is hard not to hail victory when it comes, even though it mean success for an enterprise that we have condemned as wrong. Lying in bed, I was cheered by the news and looked without regret at my wedding ring which had turned from gold into steel. It was no magic wand nor Enrico's experiments in transmutation of elements that had performed the trick. It was Mussolini's deed. The *Duce* was subtle in his choice of propaganda tools for the masses. He played on the Italians' mysticism, on their longing to believe rather than to understand, and to be made to believe through symbolism and rituals. Uniforms, parades, rallies, marches, were part of a huge choreography that had the flavor of ancient mysteries. At the time of the Ethiopian war Mussolini regained popularity by selling a new mystic idea to the people.

The war was unpopular. Nobody had wanted it. The economic sanctions had rallied the Italians around their *Duce* for a while. When the League of Nations had announced the sanctions, Mussolini had shouted at a rally:

"If sanctions come, we shall tighten our belts!"

"We shall tighten our belts!" had echoed the cheering crowd.

Soon tightened belts began to hurt, and Italian troops met no success in Africa. By late fall of 1935, fascism's popularity was at a low ebb. There was open grumbling in the streets, talk of revolution; staunch Fascists disclaimed their loyalty to the party, preparing for a non-Fascist future.

At this moment Mussolini sold his mystic idea. Knowing that a "spontaneous" sacrifice performed by the people would make acceptable the sacrifices he requested of them, he appealed to all Italian women to give their wedding rings to the Fatherland in exchange for steel ones, of their own free will, under no compulsion. For this sacrifice he devised a ritual, a pageantry of faith.

On December 18, 1935, exactly one month after the economic sanctions had become effective, women from all walks of life in all cities and villages marched in columns to exchange their rings. In Rome the endless procession of ring-bearers was led by Elena, queen of Italy. In the gray light of the early December morning the proud and stately queen paused in Piazza Venezia to face the white marble stairs up to the Altare della Patria, the Altar of the Fatherland, under which the Unknown Soldier is buried. There she placed a wreath. Then, while war widows and war mothers, aligned on higher steps, looked on her with emotion, she crossed herself, kissed her golden ring and the king's, and put them inside a crucible held by a tripod over incense-burning flames. A priest consecrated a steel ring and slipped it on her finger. All along, *carabinieri* had been playing the national anthem.

The ceremony turned out to be a genuine show of enthusiasm. It may well have saved fascism.

I had exchanged my ring like all other women, and the pride of participating in the common effort had been mingled with qualms about directing that effort to the wrong cause. But now the cries of victory, which entered the hospital room from the street, had stilled my qualms for the time being, and I could contemplate the ring, still unfamiliar on my finger, in peace of mind.

Nella had wanted a baby brother badly. From the time she had started talking, she had asked us to buy a brother. Once she had come home from a walk with her nurse in a flurry:

"Mummy," she had shouted, "give me some money right away. There is a baby in the store downstairs. I think it is for sale."

When at last little brother was to come, we tried to drive the non-sense of buying babies out of her head. Then she went around telling her friends that perhaps *they* had been bought but *she* was born from her mother.

When Giulio arrived, we did not need to give her the extra doll that experts advise to make up for that part of our attention diverted to the new baby. *He* was that doll. Motherly Nella held him in her arms, caressed his soft skull that felt silky under the hand, rocked him, helped me bathe him, and like a real mother she worried over him.

For all her chubbiness, for all the calm in her thoughtful eyes, Nella was capable of deep worries. She was excitable. Or perhaps we overtaxed her nervous system. Any psychologist could probably have warned us: to make a five-year-old try on a gas mask and to explain to her that we should be prepared for war would scare her so thoroughly that we would have to spend many hours in quieting her down. In our justification I may argue that, when danger arises, the helplessness that individuals experience under a dictatorship drives them to action. They must do something, take some positive steps, or be overcome by frustration. Hitler had occupied the Rhineland in violation of the Versailles Treaty in March, 1936. Then Enrico, aware of the ever increasing threats to European peace, had felt the compulsion of doing something for our family and had procured gas masks for all of us.

Nella was worried. She wanted to know all about wars, whether war would come to Rome and whom we would have to fight; a question not easily answered at a time when the Ethiopian war was not yet ended, when there were League of Nations sanctions against Italy, when, however, Mussolini still opposed the rising power of Germany.

Enrico and I tried to minimize wars after we saw how excited Nella was. We told her that as children would quarrel and fight now and then, so would nations, and that there was not much more than this in a war. But Nella had more sense than to believe us. If the Germans made her wear a mask, they must be dreadfully bad and to be feared. She wanted to know what they would do to her; how long she would have to keep her mask on; and whether she would be able to eat, to drink, and to sleep with that thing on her face. But, above all, she was concerned with what would befall her little brother Giulio, who was so tiny that no gas mask could possibly fit him.

If Giulio worried, he could not say so. The baby's reactions to the world were still limited. He was fond of being rocked in his bed, of being hugged by protective arms, of food, of company, and of the full daylight. As soon as he learned to smile, he went on smiling all day long and screamed all night. He disliked solitude and was drawn toward mankind. His love he reserved for his father, and with his love went the brightest of his smiles. Once—he could have been four or five months old—I entered his room with a friend who was about Enrico's build. From afar Giulio smiled at him, but when the man came closer, Giulio, conscious of his error, started crying.

As a baby Giulio was the perfect extrovert and nonintellectual. When not yet capable of standing on his feet but only of crawling on the floor, he had accomplished a feat: he had opened the dining-room cupboard, helped himself to a banana from the fruit bowl stored in there, peeled it, and stuffed it into his mouth. The peeling part of his exploit, I felt, was a sign of wisdom. But Enrico attached no importance to Giulio's deed.

The baby grew to be an alert busybody with wide-open brown eyes and attentive ears, who tried to miss nothing in this world, an eager child who approached the grownups as his peers, who listened to their conversations and mixed his baby talk with their serious words. I let him pound screws and nails into his wooden toys—all metal went into making guns—happy that from Enrico he had inherited at least manual skill. About his mental faculties I did not worry. I had learned my lesson in raising Nella. And before Giulio had begun talking coherently enough to make it possible to probe his intelligence, we left Italy and came to the United States. Busy with graver preoccupations, I gave less attention to his development than I had to Nella's.

Ten months before we left we had moved into a larger apartment. And in the spring of 1938 Nella caught the measles, and Giulio saw Mussolini and Hitler together.

This larger apartment was near Villa Borghese, the main park of Rome. We had bought it because I had been attracted by the idea of a green-marble-lined bathroom. It satisfied my ambitions of grandeur, which had been rising as Enrico's position had steadily grown better. As he had expected, the money he had never sought kept coming to Enrico: salaries from the university and from the Royal Academy of Italy; royalties from his books; a compensation for sitting on the administration board of E.I.A.R., the government-controlled broadcasting system; savings on his trips to America; interest on wise investments. During 1937, the last year we spent entirely in Italy, our total income amounted to the equivalent of $7,500, a comfortable sum in Italy. When we moved into our new apartment, early in 1938, I felt rich, well established, and firmly rooted in Rome.

Our new apartment was spacious and laid out in such a way that it was easy to quarantine Nella and separate her from the rest of the family when she caught the measles in the spring. I washed her things, clothes and dishes, in the green-marble bathroom, while Giulio was confined to the less elegant one, lined with plain white marble. For further protection I sent him out of the house as much as possible. He spent most of the day in Villa Borghese, accompanied by the nursemaid, and thus one day he ran into Hitler and Mussolini.

Hitler's visit to Italy took place in early May. Great had been the preparations. Along the road which he traveled from the north to Rome, all peasant houses had been repainted at government expense, and the Fascist slogans on them had been rewritten in blacker letters:

MUSSOLINI IS ALWAYS RIGHT.

TO WIN IS NECESSARY. BUT TO FIGHT IS MORE NECESSARY.

IT IS THE PLOW THAT TRACES THE FURROW, BUT IT IS THE SWORD THAT DEFENDS IT.

BOOK AND GUN—PERFECT FASCIST.

In the central streets of Rome, hotel and store fronts had been made over, modernized. One morning Mussolini had taken his pal for a horseback ride in Villa Borghese: children, women, and strollers had cheered, and the Fascist salute, "Eia, eia!" was mixed with the Nazi, "Heil!"

Giulio and the nursemaid had come home bubbling with excitement, and Nella and I had regretted that the measles had kept us home and deprived us of a show that would never be repeated for us.

Enrico failed to be duly impressed by the Fascist choreography and by the Fascist slogans. On one instance his disrespect for these manifestations of fascism to which I had become entirely accustomed almost shocked me. It was September, 1937, and Enrico had just returned from a visit to the United States. He had crossed the ocean with his friend Felix Bloch, a Swiss-born physicist settled in California, who was later awarded the Nobel Prize in physics.

Enrico, Bloch, and I drove together from Rome to Florence in our car, no longer the Bébé Peugeot, but a more elegant Augusta. Although the Fascist slogans had not yet been repainted for Hitler's visit, nonetheless, they were more than visible, they jumped to the eye from the faded walls of the peasant houses along the road.

My two companions, who bore in their memory fresh recollections of American advertisements, now read the Fascist slogans with an American addition, in loud, shouting voices:

Mussolini is always right; Burma Shave!

To fight is necessary, to win is more necessary; Burma Shave!

Not even these revised editions satisfied Bloch entirely. "How much better the true Burma Shave lines are!" he exclaimed regretfully, and then he quoted:

At crossing roads don't trust to luck;
The other car may be a truck.
Burma Shave.

(13)

NOVEMBER 10, 1938

A telephone that rings early in the morning has a peculiar quality. It is sudden and startling, shrill and peremptory. It shatters the silence, it fills the whole space, it reaches inside your bed covers. You are shocked out of your last dreams, you are forced out of your bed, for you cannot escape it or ignore its insistent bid. So, early on the morning of November 10, 1938, I found myself answering the telephone in the hall of our home.

"Is this Professor Fermi's residence?" asked the operator's voice.

"Yes, indeed."

"I wish to inform you that this evening at six Professor Fermi will be called on the telephone from Stockholm."

My drowsiness vanished at once. A call from Stockholm! I could guess the implications of a call from Stockholm! My slippers banged louder and louder with noisy excitement as I ran on the terrazzo floors from the hall back into our bedroom. Enrico's head was still sunk in the hollow of his soft pillow, a black spot among vast whiteness.

"Wake up, Enrico! This evening you'll be called on the telephone from Stockholm!"

Calm, but immediately alert, Enrico propped himself up on his elbow and replied:

"It must mean the Nobel Prize."

"Of course it does!"

"So the possibilities that were hinted to me have become true, and it was right to make our plans as we did."

At this reference to our plans my excitement subsided.

According to our plans we would leave Italy for good early next

115

year. But if Enrico were to be awarded the Nobel Prize, we would leave sooner, in less than a month, go to Stockholm, and then directly on to the United States, without coming back home at all.

Our plans were the most sensible under present circumstances, there was no doubt, and I had accepted them on the rational level. But emotionally I still rejected them; I still rebelled against them and against the apprehension of the unknown that the future had in store. The thought of leaving Rome gave me pain. I was born there. I had always lived there. My relatives, my friends, were there. I belonged in Rome. My roots were so firm, they reached so deeply into the rich soil of memories, habits, and affections, that I felt I would not be easily transplanted.

In past years Enrico had often suggested that we should leave Italy to escape fascism, and move to the United States; each time I had raised objections. Fascism so far had been a mild dictatorship and had not interfered with the private life of people who, like us, did not put their criticism and disapproval into action. The great majority of Italians were politically inactive; they let themselves be dragged downstream by the strong current and did not struggle against it. Perhaps it was just as well, because in a thoroughly policed and organized state such as Italy was under fascism, greater open opposition would certainly have caused more suffering but might have failed to achieve positive results. In any case, the doctrine that government is the responsibility of each and all, that every individual should participate in it, was not so widely accepted in Italy as it is in America. The scholars' ivory tower, the isolation in which they lived aloof from politics, was considered as worthy as, if not worthier than, the courageous but ineffectual acts of rebellion of some members of the intelligentsia. Despite fascism, life in Rome had been pleasant for us, and we had stayed.

Circumstances were different in 1938. The reasons for the change can be traced back to Mussolini's Ethiopian venture and the economic sanctions imposed on Italy by the League of Nations. Sanctions had been a half-measure, not strong enough to stop the war but sufficient to estrange Mussolini from the Western Powers. Embittered Italians had pursued to victory a war that had become the symbol of a fight against international persecution.

The consequences of sanctions were far more serious than a victory in Africa: they forced fascism into an alliance with Nazi Germany.

It was almost unbelievable. Germans were the traditional foes of Italians, since the first World War, a fallen foe. The newly risen Führer of Germany was held to be a none-too-intelligent imitator of the *Duce*, a puppet obediently waiting for directives from the Fascist Master. The puppet had taken some initiative of his own. In March, 1935, he had denounced the Versailles Treaty and declared that Germany would rearm. Mussolini's ire was aroused. He called a conference with France and England at Stresa and pledged himself to help contain German rearmament.

The Führer had in store another of his spring surprise moves: in March, 1936, his troops occupied the demilitarized Rhineland. By then Mussolini was on bad terms with France and England, but dreaded and opposed a strong Germany. His attitude was well expressed in a newspaper headline commenting on the Rhineland occupation:

"GERMAN VIOLATION OF THE TREATY UNANIMOUSLY RECOGNIZED," the headline said. "ITALY WILL MAINTAIN WATCHFUL RESERVE UNTIL JUSTICE WILL BE ACCORDED TO HER ON THE ETHIOPIAN QUESTION."

His bluff did not work. The "Stresa Front" broke up. In the following July, Germany and Italy found themselves together, unofficially fighting on the same side of the Spanish war.

From then on, there was avowed friendship between the two dictators. Words with the quality of lovers' smiles were exchanged, and the Rome-Berlin Axis, another symbol devised by Mussolini, came into being on October 23, 1936. The *Duce* still lived in the delusion that he was the boss and that he had Hitler well in hand. Abruptly, his delusion was shattered by the *Anschluss*. On March 12, 1938, Hitler occupied Austria without consulting or even informing Mussolini. He knew only too well that his friend would object violently: for years Mussolini had loudly played his self-assumed role of Austria's protector. When Chancellor Dollfuss had been assassinated in July, 1934, the *Duce* had sent troops to protect the Austrian border from German invasion and had shouted to the world that

"Austria is not to be touched." Germany at the Brenner Pass would be a perpetual threat to Italy.

That the *Anschluss* had come as a surprise to the Italian dictator was quite evident from the attitude, or rather lack of attitude, of the press. Whenever in the past an event of importance had taken place, the press had immediately received directives on the official point of view; on the line of thought to be followed in comments; on the amount of space to be given that event; even on the size of the head-lines. When the *Anschluss* was announced, papers and news broad-casts took no stand for several hours. There were no comments on Hitler's move, no "official interpretations." Mussolini had not decided yet whether to be publicly outraged, as he certainly was in private, and admit he had been fooled or to give his wholehearted approval and submit to the *fait accompli.* Soon the press burst out in praises for the Führer's statesmanship, for the union of two nations that always had wanted to be one. Mussolini had saved his face. But Italy had become Germany's slave.

The consequences of this enslavement were felt only too soon: in the summer of the same year, 1938, Mussolini launched an anti-Semitic campaign, for which there were no reasons, no excuses, no preparations. No real anti-Semitism existed in Italy, and Mussolini himself had so declared on several occasions. True, a few careers were more difficult for Jews to enter. True, Professor Levi-Civita, the well-known mathematician, had not been named to the Royal Academy of Italy, despite its members' repeated recommendation. True, my father had been suddenly and unaccountably dismissed from active service in the Navy and placed in the reserve. Still these were scattered incidents. There were no Jews and "Aryans," only Italians. Jews represented one per thousand of the population and were destined to decrease in number through the ever increasing rate of mixed marriages.

Shortly before our departure I overheard a man in working clothes ask another in Rome: "Now they are sending away the Jews. But who are the Jews?"

There were no Jews in southern Italy or in Sicily. From the *podestà,* the mayor, of a remote village in Sicily Mussolini was said to have received a telegram: *"Re:* Anti-Semitic campaign. Text: Send specimen so we can start campaign."

Skiing in the Alps: Fermi and Persico Are Ready To Start

Fermi Keeps Himself and His Ski Poles in Good Shape

Ann Arbor, 1930: The Dutch Group and Fermi: At the Center in the
Front Row Is Professor Ehrenfest; Third and Fourth from
Left Are George Uhlenbeck and Sam Goudsmit;
the Ladies Are Their Wives

On the Way Back from South America: Fermi
and Ottorino Respighi (1934)

No indications of a racial policy were in sight at the beginning of July, when I left Rome with the children to spend the summer in the Alps. We had rented a house in San Martino di Castrozza, one of the most spectacular resorts in the Dolomites, surrounded by the *pale* or "shovels," thin, tall, almost two-dimensional rocks, a fence of wooden shovels. Inside this fence, in the vast basin lined with green meadows, I felt the separation from the world intensely. Happy to vegetate, to relax, to watch the children become healthily tanned in the rich sunshine, I forgot fascism, naziism, and the troubles of Europe. I read no newspaper, listened to no broadcast.

August brought Enrico to San Martino. He appeared to be preoccupied, and I asked him why.

"Haven't you noticed what's going on?"

There was some surprise in his voice, but mostly disapproval, a profound disapproval that wounded my pride the more deeply for not being put into words. I would have preferred a scolding or an outburst of anger. But those I never got from Enrico.

On July 14, Enrico explained, the *Manifesto della Razza* had been published, a document which in scientific language formulated the greatest absurdities and tried to hide its contradictions in redundancy of phraseology. Separate human races exist, the manifesto stated. The Italian population is of Aryan race. Because there has not been a recent influx of masses into Italy, it can be asserted that *by now* there exists a pure Italian race. The most blatant contradiction of the manifesto concerned the Jews. It seemed as though the compilers of the manifesto found it expedient to make a distinction between Jews and Semites. The paragraph on the Jews reads as follows:

"JEWS DO NOT BELONG TO THE ITALIAN RACE—Of the Semites who through the centuries landed on the sacred soil of our country, nothing is left. Also the Arab occupation of Sicily has left nothing except a memory in a few names; anyhow, the process of assimilation was always very rapid in Italy. The Jews represent the only population that could never be assimilated in Italy, because they are constituted of non-European racial elements, absolutely different from the elements that gave origin to the Italians."

To the honor of Italians, it must be said that Mussolini had great difficulty finding university professors willing to sign this manifesto. Not a single anthropologist put his signature to the document.

The racial campaign, so brilliantly launched, acquired momentum at an amazingly fast pace. An institute was established for the "defense of the race." A magazine was published, also called the *Defense of the Race.*

At the same time the Italian government seemed to have gone crazy. New laws, rules, and directives came forth daily, at random, as if their exclusive purpose were to prove the all-powerfulness of the Fascist god. They prescribed uniforms for white-collar workers in civil service; they defined the style of women's hairdos; they banned ties from men's clothing on the grounds that the tie knot pressed on a certain nerve and prevented taking the right aim with a gun. More serious laws barred bachelors from advancement in government-controlled positions; made women's employment subject to marital status; forbade marriages between Italians and foreigners, between Aryans and Jews.

The first anti-Semitic laws were passed early in September. We at once decided to leave Italy as soon as possible. Enrico and our children were Catholics, and we could have stayed. But there is a limit to what one is willing to tolerate.

Afraid that our passports might be withdrawn if our true intentions were known, we were faced with the problem of organizing our departure in secrecy. Foreign mail was likely to be censored. Enrico wrote four letters to four American universities, in which he stated that his reasons for not accepting their previous offers had ceased to be. He dared write nothing more specific.

We were still in the Alps and four letters, all in the same handwriting, all going to America, if mailed at the same village could not fail to arouse suspicion. We took advantage of a car trip and mailed Enrico's letters at four towns miles apart.

Enrico received five offers of positions in America. He accepted that of Columbia University. To Italian officials he declared that he was embarking on a six-month visit to New York.

Then an unexpected complication came to modify our plans. In October, at a physics meeting in Copenhagen, Enrico was confidentially informed that his name had been mentioned with others for the Nobel Prize, and he was asked whether he would rather have his name temporarily withdrawn in view of the political situation and of the Italian monetary restrictions. Under normal circumstances any information concerning the Nobel Prize is strictly secret, but it was thought permissible to break the rules in this case.

To prospective emigrants, who would be allowed to take along fifty dollars apiece when leaving Italy for good, the Nobel Prize would be a godsend. However, the existing monetary laws required Italian citizens to convert any foreign holdings into lire and bring them into Italy. Hence our decision to go to Stockholm and from there directly to the United States, if Enrico were to be awarded the Nobel Prize.

Then came November 10 and the early telephone call.

"Let's celebrate," I said. "Don't go to work today. Let's go out together."

Thus a while later we were out in the streets of Rome and on a buying spree. We bought a new watch for each of us. I was proud of mine, but at the same time I felt remorse, as if I had no right to own it, as if I had come into possession of it under false pretenses.

"We have spent so much money," I said to Enrico. "Now, if that telephone call does not mean the Nobel Prize. . . . What are we going to do?"

"The probability that the call means the Nobel Prize, or part of it, in case it should be awarded to two physicists together, is at least 90 per cent. Even if it does not, we can afford a watch. Besides, we should take along a few objects when we leave. I wouldn't try to purchase diamonds because records of their sales are kept, and we don't care to have our names on records of that kind. A watch is the thing to buy. Inconspicuous and useful."

For the second time since the morning I was reminded that these were my last days in Rome. But I was determined to be of good cheer and to chase away the nostalgia that came over me at the familiar sight of the Roman streets; of the old, faded buildings that had preserved their full charm; of the clumps of ancient trees that everywhere interrupted the monotony of the streets, rising above a discolored wall or behind an iron fence, silent and monumental witnesses to human restlessness; of the numberless fountains of Rome, which indulged in the opulence of their water, shot it toward the sky, and let it come back in cascades of diamond-like droplets, in rainbow patterns. I was going to enjoy these sights and give thanks to God for thirty years of life in Rome.

At home the afternoon hours stretched out forever. Would six o'clock never come? I asked the mute telephone every time I passed it in the hall.

At a quarter of six Enrico and I sat down to wait in the living-room. It was large and comfortable. The parquet floor shone, and the woodwork gleamed. We had not lived long in this apartment; but memories pile up fast with growing children. It was surprising how much these rooms had to tell me after only ten months. There, on that piece of hard mosaic floor in the sun porch, Giulio had bumped his head so badly that he had looked like Moses for days. In the bedroom that I could see across the hall through the open door, where the rays of the late afternoon sun now fell on Nella's head bent over Giulio's while she read him a story, Nella had been ill with the measles. And in the adjoining bathroom, my beloved green-marble bathroom, I had set up the improvised steri-lizing outfit. In that corner, by the couch in the living-room, Giulio had stood with his face against the wall, in punishment for having stuffed his mouth with half a platter of French pastry prepared for company. . . .

The telephone bell rang, and I sprang up.

"I'll take it," I told Enrico and ran to the hall.

It was not Stockholm.

"Haven't you had that call yet?" Ginestra Amaldi asked. "We are waiting for the news. Rasetti is here with his mother, and the other people from the lab. Call us as soon as you have talked to Stockholm."

I sat down again. In front of my eyes the iron grayhound trapped in the lacy pattern of the radiator cover strove to go faster than time but did not move. Like the hopes of humanity . . . perhaps?

"It is six o'clock. I'll turn on the radio, and we'll listen to the news while we wait," Enrico said.

We had become accustomed to being often upset by radio an-nouncements during the last months. This time it was worse than ever.

Hard, emphatic, pitiless, the commentator's voice read the second set of racial laws. The laws issued that day limited the activities and the civil status of the Jews. Their children were excluded from public schools. Jewish teachers were dismissed. Jewish lawyers, physicians, and other professionals could practice for Jewish clients only. Many Jewish firms were dissolved. "Aryan" servants were not allowed to work for Jews or to live in their homes. Jews were to be deprived of full citizenship rights, and their passports would be withdrawn. All my relatives and several of our friends would be affected, would have to reorganize their lives, somehow. Would they succeed?

The telephone bell rang again.

"Well? What about this call? Isn't it coming?" Ginestra's impatient voice asked.

"Not yet," I answered. "But I don't know that I care for it now. Have you heard the news?"

"No, we haven't. What happened?"

"More racial laws," I said and hung up the receiver.

The telephone call from Stockholm eventually came through, and it did mean the Nobel Prize. The secretary of the Swedish Academy of Sciences read the citation over the telephone:

"To Professor Enrico Fermi of Rome for his identification of new radioactive elements produced by neutron bombardment and his discovery, made in connection with this work, of nuclear reactions effected by slow neutrons."

There was no doubt now. Enrico had been awarded the Nobel Prize. Four years of patient researches; the broken and the unbroken tubes full of beryllium powder and radon; the strenuous races along the hall of the physics building to rush element after element to the Geiger counters; the efforts to understand nuclear processes, and the many tests to prove the theories; the goldfish fountain and the paraffin blocks—these had won the Nobel Prize for Enrico.

The Nobel Prize had not been divided between Enrico and another physicist, as we thought it might. Still it gave little joy to me. I did not know whether to be happy or sad, whether to heed the telephone or the radio.

Only a few minutes later the doorbell rang. Ginestra, tall, thin, and smiling her soft smile full of sweetness, headed a line of a dozen people. Our friends, old and new, had come to congratulate Enrico.

"We are going to stay for dinner," Ginestra brazenly said, with no question or hesitation in her voice as she walked into the hall. Thus the house, so quiet and subdued a minute earlier, broke out in merry confusion, in busy commotion. The maid was instructed to set a long table, the cook consulted on how to turn our family supper into a banquet; ready-cooked food was sent for; wine was made ready for the celebration. Infected with everybody's excitement, Giulio tried to climb men's legs, to get our friends' attention, while dignified Nella vainly strove to teach him good manners.

Our celebration of the Nobel Prize was Ginestra's scheme, and a successful one, so that we should not brood over the new racial laws.

(14)

DEPARTURE

On December 6, 1938, we left Rome with our two children and their nursemaid. The journey to Stockholm was as comfortable as a train journey can be with two weary children who were soon to be eight and three years old, respectively, and who could be pleased by neither their toys nor their books.

Except for a slight incident at the German border, the only other memorable diversion to the rhythmic monotony of the train was the crossing of the rough Baltic Sea on a ferry and the shattering clangor of a pile of dishes that tumbled down from a dining-room table as the ferry hit an unusually violent wave.

The incident at the German border was so slight that it was only a brief moment of anxiety, not worth relating but for the fact that it reflected the strain and tension of our last month in Rome. We had been under the constant fear, common to all who plan to leave a country under difficult political circumstances, that we might not succeed in carrying out our projects. Something, we thought, might come up between our plans and their completion, a specific act of the government against us, a new law, the sudden closing of international frontiers, or the outbreak of war.

Enrico had never admitted he was worried. In our family his was the role of reassuring, of never having a worry or a doubt. When the order came for all Jews to surrender their passports and have their race recorded on them, I had been frightened. I foresaw a lengthy delay in the best of cases, Enrico leaving without me in the worst. But Enrico had preserved his calm and professed his confidence that everything would be well in the end, that with the help of an influential friend we would overcome this difficulty, as we had overcome others. He was right as always: within two days I had my passport back, and no record of race was on it.

Despite his reassurance and the fact that he restated it several times for no apparent reason during the first part of our journey, when the Italian guards at the Brenner Pass had examined our passports and returned them to us with no comments, Enrico had seemed relieved.

Then it was a German guard's turn to inspect our passports. He was standing in the corridor outside the door of our bedroom, stiff and official, a personification of our past and present anxieties. He turned our passports in his hands searchingly and appeared unsatisfied. Enrico rose from his seat and stood in the corridor, waiting, his thin lips so tightly pressed together that they had disappeared inside his mouth. The moment unfolded with unbearable slowness. Nella, always sensitive to our moods, became restless. Why, she wanted to know, was that gentleman taking so long with our passports? Why did he turn all the pages over and over again? Did I think something was wrong? Would the man send us back to Rome and to Mussolini?

"Be quiet, Nella. Everything is all right!"

Everything *had* to be all right. Once I had accepted the decision to leave Italy, I felt as if I had always wanted to leave; as if all my yearnings and expectations had built up over the years in one direction: America; as if failure to continue our journey now would offset a lifelong dream.

Enrico spoke to the guard in German. Was anything the matter? Had we obtained a visa from the German Konsulat? the guard asked. He did not seem to find it in our passports. When Enrico turned the pages and pointed out the visa, the stiffness vanished from his muscles, his thin lips were visible once more. The German guard saluted and smiled. Germans and Italians were good friends, were they not?

"Now, Nella, you can go back to sleep. Nothing was wrong. Now climb into the berth with Giulio and take care not to wake him up. Soon the train will start moving and will rock you."

Soon the train was in motion, forward, away from Italy.

The parting from friends and relatives had not been so hard as I had expected. I had told so often the "official version" of our trip to America—that Enrico was to teach six months at Columbia University in New York and that at the end of his engagement we would return to Rome—that I had come to believe in it. Besides, I repeated to myself, even if we were to settle in the United States, I could always come back for a visit. What could prevent me? The racial laws? There was nothing in them against my traveling. A war? Why doubt Hitler's sincerity when at Munich he had declared he had no further territorial ambitions? Why heed the pessimists rather than the optimists?

I had closed my eyes to the evident. The war was to come, and all hopes of Italy's last-minute shift in alliance were to vanish. Fascism had united with naziism, and Italy was little more than a province of Germany. Still fascism resisted total nazification. It preserved its identity in part. What was left of its individuality made it vulnerable, a designated victim of naziism. The German occupation was to result in tragedy for most Italians and in a more urgent, immediate tragedy for the Italian Jews. Some fled to hide in the Italian mountains, and some crossed the Alps on foot, into the relative security of Swiss concentration camps. They were guided by smugglers who knew the unguarded passes, who helped them carry bundles and babies, while little children who could stand on their feet had to walk endless hours. Some changed their names and lived in disguise and constant fear; and a large number, mostly the old who had felt protected by their age, were rounded up by the Germans and deported to labor camps and gas chambers.

This was to happen five years later, and, luckily for me, I had no premonition of these events at the time of our departure. My worries were mixed with a certain spirit of adventure, and a good part of my mind was absorbed in the cares of the moment.

A few people knew that we were to settle in the United States, among them the Amaldis and Rasetti. They came to say goodbye at the station in Rome, and with them we paced the platform along the train, while Nella and Giulio took possession of our two bedrooms first, scattering their toys around under their nurse's vigilant eye, and then they flattened their noses against the window to look at us.

This parting had a significance that none of us wished to put into words. It was the formal ending of a co-operation that had started almost twelve years previously. The group had dwindled down. After Segré had gone to Palermo in 1936 and while Rasetti was in the United States for a long visit, co-operation had been limited to Enrico and Edoardo Amaldi. The bulk of the experimental and theoretical work to interpret artificial radioactivity and behavior of slow neutrons was carried on by the two of them alone. But so long as an active nucleus of the group remained, there was always a possibility of restoring it to its former strength. Not so now. Emilio Segré, who had gone to the United States for the summer session at the University of California in Berkeley, had watched the trend of events in Italy. He had decided not to come back. His wife and year-old son had joined him.

Franco Rasetti was quietly looking for a position outside Europe. He was to leave Italy in July, 1939, and become professor of physics at Laval University in Quebec. Of the old group, only Edoardo Amaldi planned to stay on in Rome. The responsibility of keeping the Roman school alive rested on him, on his will power and on his abilities.

As we swiftly paced the platform up and down in the cold December morning, the comment to these unspoken thoughts came from Ginestra Amaldi. She said then what she had said before, when I had told her of our decision to leave:

"Enrico's departure is a betrayal of the young people who have come to study with him and who have trusted in him for guidance and help."

"No, you are not being fair," Edoardo remonstrated. "Enrico honestly meant to fulfil his duties toward his students. He would not have left them without timely warning, had circumstances been normal. His reasons for leaving the country are impelling and independent of his will. Fascism is to be blamed, not Fermi."

Ginestra shook her head, and her expression became stubborn, as only the expression of usually yielding and gentle people may become. Her words were an incomplete formulation of her thoughts. Questions were in her mind that have troubled humanity since man started his quest for rules of conduct valid under all circumstances. Of contradicting duties, which should one choose?

Should the responsibilities toward one's family or those toward one's students come first? Should love of one's country come before love of one's children? Should one forgo the opportunity to take one's family to security and a more suitable environment in which to raise children, or should one remain under a despised government waiting for a possible chance of helping fellow-citizens from within? And, perhaps the most perturbing of all moral contradictions, should a woman forget her duties as a daughter to follow the call of those as wife and mother?

I knew Ginestra well enough, her deep devotion to her parents, her unshakable religious faith, to be able to read all this in her stubborn eyes. Edoardo had also read her thoughts, and his words had been an attempt at balancing her emotional convictions with his own rationalism. But questions that have not found an answer over the centuries cannot be settled in the few minutes before a train departs.

Ginestra was still stubborn, and I was full of doubts, when the railroad man shouted: "All aboard."

"I hope I'll see you soon," Rasetti said, and his voice was more subdued than I had ever heard it.

We climbed inside the train, lowered a window, and leaned out to wave goodbye to our friends one last time, as the train whistled and shook itself into sudden motion. When they had faded away, we closed the window against the sharp air of the winter morning. I took off and carefully laid down my new beaver coat—part of the refugee's trousseau, in Enrico's words, that we were taking along instead of money—and slumped in my seat.

The aqueducts and the umbrella pines of the Roman countryside sped by.

"Nothing can stop us now," Enrico said.

Now we were part of a moving train, of an inflexible schedule that would refuse to hesitate at frontiers and in forty-eight hours would bring us to Stockholm.

The fallacy in this reasoning is that trains have doors, and through those doors people can be made to get off. Neither Enrico nor I dared to mention this fallacy, although we were both aware of it. It conferred infinite power upon guards and customs officials; it distorted the significance of their actions; and it transformed the minutes they spent over our passports into bits of eternity.

Luckily, that sort of eternity moves along with time. It moved along with our train. We came out of Italy and Germany, and we could relax. We could enjoy the luxury of our first-class accommodations, of traveling with a maid along, who could find little more to do than roll Giulio's big curl on her finger to shape it in a long tube on top of his head.

That she could come with us was due to the Nobel Prize. When in October I had expressed the wish to have her with me during the first months in New York, Enrico had gone to sound out the American consul, who had been very helpful so far. The consul had not been encouraging: the Italian quota was filled up, and there was no hope of an immigrant's visa for the maid. As for a visitor's visa, what assurance could she offer that she would actually be a visitor, that she would return to Italy and not outstay her permit? We almost gave up. Then the Nobel Prize came and produced a blossom of smiles inside the consulate. The fiancé whom our maid produced was considered a sufficient guaranty, almost a hostage, of her coming back. In a few days she had her visa.

The Nobel Prize achieved other feats at the American consulate. When the American physician who examined us found that Nella used her right eye only and could see little from the left, he was inclined to raise serious difficulties: American health standards were to be kept high; Nella's vision defect ought to be corrected before we should be granted permission to enter the States, he contended. But the words "Nobel Prize" whispered into his ear silenced his objections.

The Nobel Prize, powerful as it was, did not exempt Enrico from taking the required arithmetic examination, which was considered as some sort of rough intelligence test. A woman came into the doctor's office, where candidates for immigrants' visas were waiting, and questioned them all.

"How much is 15 plus 27?" she asked Enrico.

Deliberately and with pride he answered "42."

"How much is 29 divided by 2?" "14.5" Enrico said. Satisfied that his mind was sound, the woman went on to quiz the next candidate. Giulio was too young for arithmetic. Nella and I passed the test. But the family of a retarded ten-year-old girl who could not keep her figures straight was denied a visa they had dreamed of for years.

At last our train arrived in Stockholm. We made our children wear the first leggings of their lives against the rigors of the northern climate, and we got off the train. Then we were dragged into the vortex of the Nobel celebrations.

There was the prize award on December 10, the anniversary of Nobel's death. Only the prizes for literature and for physics were awarded in 1938. Pearl Buck, the American writer of novels with Chinese background, and Enrico sat in the center of the stage in the Concert Hall. The hall was filled to capacity with bejeweled women in low-necked gowns and solemn men wearing white ties, tails, and heavy decorations on colored ribbons. Behind Pearl Buck and Enrico were past years' recipients of Nobel Prizes and members of the Swedish Academy.

Attentive and stiff in their tall armchairs with embossed leather backs and carved lion heads, Pearl Buck and Enrico faced the audience while listening to music and speeches. Plump and attractive in a soft evening dress, the train of which she had gracefully gathered around her, a pensive smile on her pleasant face, her hands demurely resting on her lap, Pearl Buck sat still; her stiffness was the outward projection of her bewilderment at the undemocratic manifestations of an Old World order and of her astonishment at being the object of such a manifestation.

Enrico sat stiff because he could not do otherwise. Stiff with the expectation of a dreaded but likely mishap: that the heavily starched front of his evening shirt might suddenly snap and thrust out in a protruding arc between the silk lapels of his full dress suit, with an explosive sound, at his first incautious move, as it had done many times before. Although dedicated to measurement, Enrico had not yet recognized that fronts of ready-made shirts were too long for him.

The Nobel medals and diplomas were given to Pearl Buck and Enrico by King Gustavus V of Sweden, who got up from his seat in the center of the first row on the floor but did not climb onto the stage. He waited for the two recipients to come, one at a time, across the stage and down the four steps to the floor. Tall and thin, he bent down over them his ascetic face, in which the skin had the pallor and the transparency that makes one wonder whether the blood of old aristocrats is not truly blue.

His Majesty shook hands with Enrico, when his turn had come, and handed him a case with a medal, a diploma, and an envelope. ("I think," Nella said later, in her calm, speculative tone, "that the envelope is the most important of the three because it must contain the money.")

With the three objects in his hands, Enrico retraced his paces backward, up the four steps and across the stage, because to royalties you must never turn your back. So, without even looking over his shoulder, outwardly sure of himself, he found his way to his leather-backed chair and happily dropped into it. Of this feat he was to brag for years to come!

And there was the night when I danced with the prince.

"The night when I danced with the prince" was the name of a popular perfume in Italy. It was intended to appeal to young romantic girls in whose dreams a charming prince often comes and asks them to dance.

If ever as a romantic girl I had dreamed forgotten dreams of this sort, not even in the dim imagery floating out of the mind in the night did I raise my hopes to a Crown Prince who was to ascend a throne within my lifetime. Crown Prince Gustavus Adolphus, now King Gustavus VI, was fifty-six years old in 1938, as dark and as robust a man as the king his father was white and frail. With him I danced the Lambeth Walk in the marbled splendor of the Town Hall. I had never danced it before, but the Crown Prince was a good leader. He was a man who gave support in a dance and inspired confidence, who wore honest, round, dark-rimmed eyeglasses à la Harold Lloyd; a man with the solid attributes of the good human being, not the unreal creature in the dreams of an incorrigibly romantic girl.

And there was the king's dinner at his palace, with a galaxy of princes and princesses, of court dignitaries and ladies-in-waiting; ladies, who, like all other women, examined the material of my evening dress—also part of my "refugee's trousseau"—and asked where I had bought it and who had made it into a gown.

King Gustavus V was my second king, and his dinner party my second dinner at a king's. I had eaten my first royal dinner five years before, with my first king, Albert of Belgium, the mountain climber.

It was in October, 1933, and we were in Brussels for the Solvay meeting of physicists. At all previous Solvay meetings the senior or the most prominent physicist from each country and his wife were asked to the Royal Palace. In 1933 Enrico and I were included in the invitation, for Enrico was the only physicist from Italy.

A short before-dinner conversation with the queen, during which my foremost concern had been to abide by the rule of never saying No to Her Majesty, had proved inconclusive and unsatisfactory because I felt I was behaving wrongly. Then we had gone in for dinner. At the place of honor by King Albert sat Marie Curie, twice a Nobel Prize winner, for chemistry and for physics. She was past middle age and had the aloof and absorbed expression of one used to taxing her intellectual powers at all times. Seated beside her, the mountaineer king seemed relaxed and affable. He slumped down and sought comfort in his big chair, he laid his strong arms on the table, next to the golden plate in front of him. The center of the table was set with golden plates, but they stopped short of me, and I missed my unique chance to eat off gold.

King Albert was a hearty eater. When fruit was passed to him in a beautiful basket, he took a pear in his large hand, and with a knife he peeled and quartered it, still holding it in his hand. My mother had taught me to lay my fruit down on my dish, to hold it firmly there with the fork, and to carve the peel off with skilled use of the knife. My mother claimed this was the only acceptable way of peeling fruit in good society.

Although twenty-six years old and a married woman five years, I exulted while watching the king and his pear and envisioned my future triumph over my mother.

Unlike King Albert's, King Gustavus' appetite was poor. Of this we should have been warned; as soon as he laid down his fork and knife after just a bite or two, the waiters, who had stood a few steps behind the sitting royalty and guests, sprang forward to the assault, grabbed and took away all plates, and served the next course, only again to remove the plates that there was no time to empty, as soon as the ascetic king had had enough.

Of all the princes and princesses who sat at the king's table, the most charming, the most softly human, was Sybille, the wife of the king's grandson. She talked to me at length in the Town Hall. Her words flowed out of her pretty mouth with easy friendliness as she inquired after my children, as she spoke of her three little daughters with motherly pride. (Later she had one more daughter and a son.) Despite the glistening diadem on her head, despite the big pearls on her white décolletage, I forgot that she might be a queen one day and addressed her as I would a newly acquired friend. Perhaps there was something prophetic in my attitude: Sybille will never become a queen, because her husband, Prince Gustavus Adolphus, who might have become Gustavus VII, was killed in an airplane crash in 1947. His and Sybille's son, now heir apparent Carl Gustavus, was then one year old, and Sybille was left with a task more difficult than preparing to become a queen—that of raising a future king.

A motion picture taken in the Concert Hall in Stockholm during the ceremony of the Nobel awards was soon thereafter shown in various countries. It produced a wave of criticism in Italy.

In 1938 the award of the Nobel prize to an Italian was viewed with manifest doubt, almost with fear, in official quarters, as something which might displease the intransigent northern ally. Hitler had prohibited Germans from henceforth accepting the Swedish award when, in 1935, the Nobel Peace Prize was conferred on Carl von Ossietzky, author and pacifist, while he was detained in a Nazi prison on charges of being an enemy of the state.

Proof of the press uncertainty was the fact that Pearl Buck and Fermi had equally shared an exceedingly brief, three-line announcement in the Italian papers. Subsequently, Fermi had gone to Stockholm and had committed a double crime: first he had failed to give the Fascist salute to the king of Sweden; then he had shaken hands with the king, a gesture that had been condemned in Italy as un-Roman and unmanly.

This wilful restraint on the most harmless of human actions seems a joke now that fascism is long dead and done for. The tragedy of Italy lay not in the fact that of such jokes there were many, but in the fact that a large number of people took them seriously. Enrico's handshake with King Gustavus was taken seriously.

Karl Sandels

At the Nobel Ceremony with Pearl Buck: Will the Shirt Front Snap Out?

Karl Sandels

King Gustavus V of Sweden Presents the Nobel Prize to Fermi

The newspaper *Lavoro Fascista* published a column-long story. It was the writer's account of his personal experience. One evening, he said, wishing to entertain a German colleague, he took him to the movies. When the time came for the *Giornale Luce,* the Italian newsreel, there appeared on the screen the king of Sweden and Fermi, both wearing not a uniform but the *bourgeois* tails. The Italian newsman felt uncomfortable. So uncomfortable that his German guest felt it his duty to put him at ease.

"If I am not mistaken," he said by way of a friendly opening, "that is your youngest academician."

"No, he is not young," the Italian answered, complacent in his promptness and wit. "In fact he is very old. So old that he is not able to stretch out his arm."

The German understood and smiled.

The behavior of most of the Fascist diplomats outside Italy contrasted greatly with this pettiness at home.

We met in Stockholm the Italian minister to Sweden, by necessity a Fascist, by birth a member of that aristocracy that, depending on a salary for survival, had traditionally filled the high ranks of Italian diplomacy. His profession had kept him in foreign countries, away from the daily indoctrination and pressure of fascism. He had been able to look at fascism from a vantage point and obtain the right perspective, like a spectator in a theater who observes fictitious happenings through well-focused opera glasses.

Although he understandably avoided discussing Italian policies with us, his whole attitude was broad, sensible, unafraid. He was more hospitable toward us than his position required, and in his hospitality he brought personal warmth. He consciously discarded from his mind the likely criticism he might be calling on himself, for he could not fail to know that Enrico was in no odor of sanctity at home, with a Jewish wife, the Nobel Prize, and this trip to America which nobody truly believed would last only six months.

PART II

AMERICA

(15)

THE PROCESS OF AMERICANIZATION

Wake up and dress. We have almost arrived. The children are already on deck."

In reluctant obedience to Enrico's peremptory voice, I emerged out of sleep and the warm comfort of my berth. It was the morning of January 2, 1939, and the "Franconia" was rolling placidly, with no hurry or emotion, bringing to its end a calm voyage.

On deck Nella and Giulio rushed to me, away from the watchful presence of their nurse.

"Land," Nella shouted; and Giulio extended a chubby finger in the direction of the ship's bow and repeated: "Land!"

Soon the New York skyline appeared in the gray sky, dim at first, then sharply jagged, and the Statue of Liberty moved toward us, a cold, huge woman of metal, who had no message yet to give me.

But Enrico said, as a smile lit his face tanned by the sea:

"We have founded the American branch of the Fermi family."

I turned my eyes down to examine my children. They seemed more thoroughly scrubbed and polished than children I had seen in America. Their tailor-made coats and light-gray leggings were different from those of other children on the boat. On their curly heads the leather helmets we had bought in Denmark against the first northern rigors appeared alien. I looked at Enrico and at his markedly Mediterranean features, in which I could read the pride and the relief of one who has satisfactorily guided his expedition across land and sea, bearing all the responsibilities on his broad shoulders with an imperturbability that would have long been thrown off had it not been so deeply rooted in him. And I looked at the maid who had come along with us, now bravely winking against the wind and rubbing her hands together to make up for the slight warmth of the coat she was wearing that had been mine, who could talk to none but us because she knew no English at all.

"This is no American family," I thought to myself. "Not yet."

But we were already undergoing the process of Americanization. It had started ten days before, shortly after we had boarded the "Franconia" in Southampton, on December 24. The children and I were exploring the boat. We found ourselves in the gymnasium, on the lowest deck, and, having decided to take a walk on the upper promenade, we called the elevator. As its doors swung open, we were face to face with a short old man in a baggy red suit and furry white trimmings, with a long white beard and twinkling blue eyes. The three of us stood still, fascinated, open-mouthed. The queer old man motioned us inside the elevator and then, with a benevolent smile, said to us:

"Don't you know me? I am Santa Claus."

Of course, I should have known him from my English teacher's tales of many years ago and from illustrations in English books for children. Still bewildered by my first encounter with a true Santa, I kept open-mouthed and had no words to say.

"I hope you'll be coming to my party this evening. I'll have presents for you!" Santa Claus said, bending his white beard toward my children. Their eyes sparkled. They turned to me: "Will you let us go? Please let us go!"

"Of course you may go. And thank you kindly, Sir."

Later I tried to explain Santa Claus to the children. Giulio, of course, could understand little of what I said, but his eager eyes were wide-open, attentive, as they always were when grownups were speaking.

"In each country of the world," I told them, "once a year children receive presents from a person who is not one of their parents, who comes for the sole purpose of bringing toys and candies."

"The Epiphany!" Nella interrupted.

"Yes, in Italy it is the Epiphany who comes on the sixth of January, the day the three kings brought their presents to the Child Jesus. She rides on a broom in the sky. . . ."

"... although she is so old one can't understand how she does it," Nella put in.

"Brings toys to me, too," Giulio said. Nella turned to him:

"She has a big, big sack on her shoulders," she explained, "and when all children are asleep and it is night, she comes down the chimney, or, if there is no chimney, she enters the door and stuffs the children's stockings with toys."

"For me too," Giulio said.

"This happens in Italy. But in America there is Santa Claus. He does not ride a broomstick but a sleigh pulled by reindeer, which are animals with big antlers. So he travels more comfortably and can carry a larger bag of toys. He comes once a year, the day before Christmas."

"Will the Epiphany come to us all the same? She knows we are Italian children. . . ."

"No, she will not. She could not get a visa and must remain in Italy," I answered on the inspiration of the moment.

"Poor Epiphany," Nella said wistfully, "I don't think she likes Mussolini too well."

Thus, of our own will, we had already accepted the first switch of traditions when the "Franconia" passed the mute Statue of Liberty and entered the harbor of New York.

For six months we lived in New York City, within the ten blocks between One Hundred Tenth and One Hundred Twentieth streets, where most Columbia University teachers live, one of the many villages into which the big city is divided. As in many small villages, there also one meets one's acquaintances in the streets and stops to greet them and to exchange some gossip. I seldom needed to go out of the village, for within it I found all the necessities of life; I came to know each street, each corner, each shop of the village: the Chinese laundry and the small tearooms, the bookstore and the post office, the College shop for men and the ladies' ready-to-wear. Only occasionally did I take a trip downtown, and it was an expedition comparable to that of the villager going to town. So I ignored the fact that New York is a huge city, and I did not feel lost in it.

For the first few weeks we stayed at the King's Crown Hotel, an old, well-established hotel in the ten-block village around the university. Then we rented a furnished apartment from a Mrs. Smith. I called her Mrs. Zmeeth and produced the blankest of expressions on the elevator boy's face each time I asked for her. The apartment was on Riverside Drive with the view of the majestic Hudson River and the placid traffic upon it. At night the Palisades glittered with many little lights beyond and across the strip of darkness that the river had become. It was a pleasant apartment. But when I led my children into One Hundred Sixteenth Street, funnel-shaped between two round-cornered buildings at its opening into Riverside Drive, the icy winter winds, which gathered violence and impact in that funnel, made us stagger, and even threw little Giulio onto the ground. Each time he walked that way, he protested and rebelled and had to be dragged by the hand, while huge tears streaked his reddened, babyish face.

That winter Giulio went daily to the playground on the Columbia campus. There he tried to teach Italian to his teachers and proved impervious to the English language. He played quietly by himself, but his big brown eyes did not miss the pleasant sight of a lithe little girl with hair the color of straw and eyes as blue as a piece of sky. She was the first typically Anglo-Saxon little girl he had had leisure to watch and his first sweetheart. He was three years old. And because he refused to learn English, she called him "the little boy who cannot talk"; but both could smile.

On a friend's advice Nella went to Horace Mann School. It was called a progressive school, but I did not know the meaning of the word and did not worry. Nella was placed in third grade, which she had started in Italy, and for a few weeks could not understand what was going on around her. As soon as she caught up with the language, I went to inquire of her teacher about her scholastic achievements. I was told that those did not matter, that the only important thing was that she should be socially adjusted. By the end of the school year I realized that Nella had conscientiously done all of her work except arithmetic. The school did not require any work she did not care to do because of her language difficulties, and she had done all that needed knowledge of the language, omitting only the universally symbolized arithmetic. She was also given an intelligence test at Horace Mann, in which, I was told, she did quite well as a whole, although she failed to answer one of the simplest questions: A little boy has taken a hike in the country where he has played with a small animal. Upon returning home he must wash up thoroughly and change his smelly clothes. What is the animal he has played with? The fact that there are no skunks in Europe or in European children's tales proves the dependence of intelligence tests on environment and vocabulary.

Meanwhile, the nursemaid, now general houseworker, and I had joined forces by tacit agreement to confront and overcome the difficulties of an American ménage. We cooked together. I had never cooked before, but I could hold the American cookbook, translate quantities into sensible metric measures and interpret directions on cans, while she worked deftly with the mixing spoon and the salt shaker. I had no idea how much salt was needed and I spent two hours salting a soup the first time I had to do it alone! Together we made fun of American recipes, in which the main concern is wholesomeness and avoidance of anything rich, rather than taste and giving pleasure to the palate.

For most of the gadgets in the Smiths' apartment I found an explanation, but the refrigerator puzzled me a long time. It was evident that it stayed cold by itself with no need to turn a switch or press a button. But neither the maid nor I could ever foresee the moment when out of deep silence it would suddenly come to life and startle us with its loud buzzing. We stood in questioning watch of the cold white bulk, which was no more willing to yield its secret than the Sphinx of Egypt. Sometimes banging its door seemed to put it in motion, but at other times it resisted the most violent slammings. A more temperamental creature I had never met.

Shopping was a co-operative enterprise shared by the maid and me. She could judge the quality of fruit and vegetables, recognize the cuts of meats. I could better translate dollars into lire to decide whether prices were reasonable; I could explore packages and cans, of which I bought large quantities, for, like any newly arrived European, we went on a canned-food spree which was to last only as long as there were new cans to try. I patronized the small shops where the clerks could take the time to instruct ignorant foreigners in the marvels of pudding powders and of the frozen foods which had just appeared on the market. In almost every grocery store one man at least was Italian born or of Italian descent, and with him my maid and I made friends at once. Not that it helped much: Italians in New York come from the south of Italy, and they bring to their speech so much of their Neapolitan or Sicilian dialects that it is hard to understand them, whether they speak Italian or English.

The maid and I used to slow down our pace while walking by the large market on Broadway, near One Hundred Fifteenth Street. We peeked in with curiosity but dared not enter; how could we have found our way in the midst of that steady agitation of women and shopping bags, of clerks and weighing scales, which filled the little space left in the mysterious pattern of displayed food? Self-service markets were still rare, and there were none in our neighborhood, or I would have become an addict, as I soon became addicted to dimestores and mail-order houses. There I could obtain what I wanted without talking, even buttons and dress patterns and all the other objects with the unpronounceable double *t* in them. My incapacity for properly pronouncing double *t*'s outlived all other language difficulties. Months later, at a time when I could usually make myself understood and had mustered enough courage to do an occasional bit of shopping over the telephone, I once ordered butter and received bird seed. We never had a canary, and the small unused package followed me across the country, a dusty reminder of difficult times.

After six months in America our maid was due to return to Italy, but she had meanwhile danced with a gentleman in a tuxedo. This dance was a symbol of fallen barriers between the classes, and after it she could no longer accept her lot and go back to a fiancé who would never give her such a social thrill. So she stayed.

Among the several traits that make a strong individualist of Enrico, one is most pronounced: the intolerance of living in a home that he does not own. Accordingly, as soon as we had settled in the furnished apartment that we had rented for six months, we tackled the problem of buying a permanent place in which to live.

It had been simple to buy an apartment in Rome: we had looked at the advertisements in the papers, we had explored a limited number of possible apartments, and we had bought one. In New York it was different. New York was a huge city, but university families were crowded into a remarkably small section of it. In that section there were no small homes, no co-operative apartments, nothing we could buy. Those of our friends who owned their homes lived in the suburbs and practiced that un-European activity, commuting. There were myriads of suburbs, but there was no Greater New York general real estate agent, who could explain the pros and cons of each place and give us an over-all picture of living conditions.

"Several of my colleagues live in a town called Leonia. It is in New Jersey, just across the George Washington Bridge, on the other side of the Palisades," Enrico said one Sunday. "Let's go see what it looks like." It was February, and an icy-cold afternoon. As we got off the bus at the stoplight in Leonia, a gust of wind blew in our faces and blinded us. We did not know where to go.

"Harold Urey, the chemist and Nobel Prize winner for 1934, lives here. We may go visit him and his wife. I know him well enough." This last sentence of Enrico's was an answer to my doubtful expression.

The Ureys were in their large living-room and had a fire going. Our visit was a success. Frieda and Harold Urey were friendly. Bashful, from a distance, their three little girls stared at us with open mouths and rounded eyes. Dr. Urey talked at length to us, in his serious, slightly professorial tone, about Leonia and its excellent public schools, about the advantages of living in a middle-class town where one's children may have all that other children have. He smiled often, but his smile stayed on the surface, as if superimposed over a serious nature. On his round face, which had just begun to become lined, there was purpose and deep concentration, almost constant concern.

Harold Urey was a good orator and sold Leonia to us. Besides, I was anxious to go live where the dirt on my children's knees would not be gray, as in New York, but an honest brown.

By the following summer we were the happy owners of a house on the Palisades, with a large lawn, a small pond, and a lot of dampness in the basement. By the time the house had been redecorated for us and was ready for occupancy, our furniture had arrived from Italy and war had broken out in Europe. We settled, and this time for good, or so we thought.

Neither of us had ever gardened. Enrico is a product of apartment buildings, and I spent most of my childhood and youth in a house whose garden was intrusted to a gardener, except for a peach tree that was under my grandmother's special care.

My grandmother was an intelligent woman, with gray hair neatly parted in the middle and brushed back in tidy bands over her temples. She had had only one daughter, my mother, and so she lived with us. I remember her a few years before she died at seventy, matronly and heavy-set but straight and dignified, clad in the black dress of a widow even when gardening, when with no ease and with sluggish motions she climbed the stepladder against the peach tree to perch on it and carefully remove diseased leaves and buds. She tried to explain to us children the basic rules of gardening, but if I listened to her out of deference for her status in the family, I have since forgotten what I unwillingly learned.

So neither Enrico nor I was qualified to take care of the lawn and the flower beds and the rock garden around the tiny pond. But we wanted to become genuine Americans and were going to do all that others did.

"On Sunday," Harold Urey had told us, "you put on your worst clothes and garden."

I was not concerned about the work in the yard and trusted that Enrico would do it. When we had married, he had told me his plans for the future. He was going to retire at forty. No physicist ever accomplishes anything after forty anyhow. He was of peasant stock and would go back to the soil. The farmer's lot appealed to the individualist in him: a farmer is his own boss and self-sufficient, for he can produce almost all he needs. Enrico had long set his eyes on a piece of land on Monte Mario, a hill on the western outskirts of Rome, with a view of the Eternal City and of the dome of Saint Peter's in the foreground.

When in Enrico's thirty-eighth year of age we settled in Leonia, his peasant blood was not aroused. Whenever the lawn needed mowing Enrico had urgent work to do at the laboratory, even though it was Sunday. By the time he could be persuaded to do the job, the lawn had grown so wild it was impossible to mow it. When it was time to water grass and flowers, Enrico preferred to go for a walk, or to play a game of tennis, claiming that the sprinkling could wait.

So I did my best, gathered all possible advice, filled our flower beds at random as our friends thinned theirs, dug the ground with a cultivator in my right hand and a garden-book in my left. The lawn did not thrive.

"The trouble with lawns around here," Harold Urey said, with serious concern, "is crab grass. You must fight crab grass. Always. Never relent. Walk on your lawn with your eyes on the grass, and when you seen one single strand of crab grass pull it out. Don't give up." He spoke with authority, and each sentence cut the air with the impetus of an ax against a tree.

By the next spring our family was all set to undertake crab-grass extermination. But which was crab grass? We pulled out the most likely plant and dispatched Nella to the Ureys' with it.

"It's not crab grass. Mr. Urey says it can't be. It's too early in the season," Nella reported.

Summer came, and still we did not know which was the crab grass. Harold Urey came by one day. He looked at our lawn, and I saw an intensification of his steady concern in his kindly eyes. He turned to me, and in a soft voice meant to lessen the impact of the news, he said:

"D'you know what's wrong with your lawn, Laura? It's *all* crab grass."

It was the summer of 1940. The phony war had long ended, and France had fallen. With an intensification of concern even stronger than that with which he had viewed our lawn, Harold Urey used to talk of the dangers of war for the United States.

"Would you be surprised," he asked his friends, "if the Germans should land at Nantucket Island by Christmas?"

During the war questions of this kind were asked everywhere and at all times. In the spring of 1941 Enrico and a few other professors at Columbia University organized a "Society of Prophets." On the first day of each month, during the lunch hour at the Men's Faculty Club, society members wrote down ten "Yes or No" questions about events likely to occur during that month. Would Hitler attempt to land in England? Would an American convoy be attacked by German ships in violation of United States neutrality? Would the British be able to hold Tobruk? The "Prophets" wrote down their answers. These were checked on the last day of the month. Records were kept of each Prophet's score.

By the time the society dissolved, Enrico had the highest score, and was *the* Prophet. Ninety-seven per cent of his predictions had come true. In foreseeing events Enrico was helped by his conservatism: he maintains that situations do not change as fast as people expect. Accordingly, Enrico had predicted no changes: Hitler would not attempt a landing in England during the month considered; the British would hold Tobruk; no American convoy would be attacked. His conservatism made him foresee no German attack on Russia during the month of June. Thus he missed a perfect prophet's score.

Meanwhile Harold Urey gardened, and Enrico tried to gain theoretical knowledge about gardening.

"Why are you so concerned about crab grass? It's green, and it covers the lawn. You people are always fighting weeds. What distinguishes weeds from other plants?"

"Weeds grow spontaneously, without being planted," Dr. Urey answered. "They take up space, air, and food from good plants and kill them. At the end of the season they die and nothing is left."

"Therefore, a weed is an unlicensed annual," Enrico concluded, following his need of defining a concept before accepting it.

If Enrico was not helpful in the garden, he was, or at least tried to be, helpful in the house. He realized that with one maid instead of the two we had had in Italy and a house instead of an apartment, there would be more work to do. To help he took up polishing his own shoes. The maid looked at him with disdain, and after several days she reported to me:

"The *professore* polishes only the front of his shoes, not the part over the heels."

Confronted with this accusation, Enrico pleaded guilty. He could not be bothered with the half of his shoes that *he* could not see.

Enrico had manual skill and learned to do the repairs around the house, like a true American husband. When using his hands, Enrico enjoys the novelty of what he is doing but stops a long way this side of refinement, as soon as the functional scope is achieved. Once, when we were still in Italy, our friend Gina Castelnuovo, the mathematician's daughter, sprained an ankle on a mountain hike in the Alps. She had to walk several miles to reach the village and a physiologist, the closest there was to a physician in the small colony of scientists gathered around the Castelnuovos. Her ankle was very swollen. The next day Enrico went to inquire after her condition. He found that the physiologist had put her leg in a cast over the swelling, that the swelling had now subsided, and that Gina's cast felt uncomfortable. It was a challenge for Enrico. Under Professor Castelnuovo's skeptical eyes, he undertook to dampen, tear apart, and reshape the old cast.

"The cast," Gina recalled later, "looked awful, but felt good."

Now in a country where the price of labor advised him to exploit his manual dexterity, Enrico set to work with that same disregard for refinement, with which he had made over Gina's cast. Did our dining-room table need extension leaves because those belonging to it had not arrived from Italy? Very good, Enrico would make them. But they would be rough and unpainted, they would always have to be hidden under a tablecloth and serve a purpose that had nothing to do with aesthetics. Were friends gathering pieces of furniture for their homes? All right! Enrico would build a rocker for them. A rocker is an Anglo-Saxon piece of furniture, not Italian; a new problem. Enrico built it, but never took the time to correct the seat inclination, which kept the sitter at an acute angle, as if bent in pain. The rocker rocked, Enrico contended, and what more can one ask of a rocker?

As soon as the challenge of a new problem was met, Enrico gave up. After all, he is a theoretical physicist and not interested in his projects, once they outlive their theoretical appeal. One set of objects never lost its appeal for Enrico: the gadgets. To him they represent the never ending quest for saving labor, the material proof of human progress, the product of a technology which he considers the symbol, the salvation, and the promise of America. He has never lost interest in gadgets, and, although parsimonious by nature and education, he is always ready to buy one more: from the step-on garbage can, his never-forgiven present of my first Christmas in the United States, through electric razor and electric saw, to the lately acquired television set, we have gone through purchase and use of all available and most automatic household equipment.

In learning the American language and habits, Enrico had a considerable advantage over me: he spent his days at Columbia University among Americans, and inside the very physics building he found an obliging mentor in Herbert Anderson, a graduate student who planned to work for his Ph.D. under Enrico's guidance.

No day went by without Enrico's telling me something that Herbert Anderson had taught him.

"Anderson says we should hire our neighbors' children and pay them a penny for each of our English mistakes they correct. He says it is the only way of learning the language efficiently."

"Anderson says"—and this it was very difficult to believe—"that English words should be pronounced with two accents: vocab'-ula'ry." It sounded very unnatural indeed.

"Anderson says that students work their way through college by waiting on tables and selling newspapers. I am afraid they have little time left for study."

"Anderson says that there are no oral examinations in American universities. The multiple-choice tests, Anderson says. . . ."

Altogether, Anderson appeared to be a bottomless well of information, and, duly impressed, I pictured him in my mind as a ponderous person, more mature and professorial than his years. But when I met him I was forced to change my ideas. He was of medium height, as slender as a boy on the threshold of manhood, dressed with the elegance of the young bachelor fond of clothes. Under well-trimmed chestnut hair, his features were small and quiet and his manners unobtrusive. But he was not a self-effacing man, and will power gave strength to his light frame.

Enrico and Anderson were fond of each other. Some young people are occasionally shy of Enrico. Some students complain that he does not know how to give them a "pat on the back." Anderson had no place for shyness and felt no need of special encouragement. Had I been able to understand Americans at that time, I would have recognized in him at least one attribute of the Jeffersonian heritage; the inborn conviction that men are created equal. The older men's position, the public recognition they might receive, the honors shed on them, were to him only indications of the goods available to mankind, and of these Anderson wanted to get his share. So Anderson was not only Enrico's student but his friend and his teacher. He learned physics from Fermi and taught him Americana.

I stayed home most of the time, got Anderson's teachings only at second hand, and learning English was a very slow process.

One day Nella came to me and said in a stern voice:

"Mother, Giulio uses bad language. I heard him call his friend *stinky.*"

I could not reply to her, not knowing the meaning of the word. I asked Enrico when he came home.

"As far as I know, it means 'having a bad odor,' " Enrico said. "But I'll ask Anderson in the morning."

From Herbert we had our first authoritative lesson in bad language: *Lousy* is not so bad as *stinky*, Herbert Anderson said. While *gosh* is almost charming in a child's mouth, *golly* is objectionable, and anything stronger is to be frowned upon. *Jerk* and *squirt* are terms which at that time children in the intermediate grades used to designate the teachers they disliked.

Nella and Giulio made me ponder not only on language but also on social philosophy. I started to understand the meaning of democracy and its institutions when nine-year-old Nella asked for "more freedom" and implied I was infringing upon her rights because I requested that she come home after school before going to play and that she let me know at all times where I could find her.

When four-year-old Giulio, whom I had asked to go wash his hands, answered "You can't make me, this is a free country," then we learned some more. To these days Enrico has retained the childish expression "this is a free country" that he acquired from Giulio, who in his turn has entirely outgrown it.

Long would be the list of what we learned from our children, besides bad and good language, the spirit of independence, and the firm belief in human rights. Looking through their young eyes, not dimmed by visions of Old World traditions, we acquired a fresh, if vicarious, perspective of American habits and viewpoints.

In the process of Americanization, however, there is more than learning language and customs and setting one's self to do whatever Americans can do. There is more than understanding the living institutions, the pattern of schools, the social and political trends. There is the absorbing of the background. The ability to evoke visions of covered wagons, to see the clouds of dust behind them in the golden deserts of the West, to hear the sound of thumping hoofs and jolting wheels over a mountain pass. The power to relive a miner's excitement in his boom town in Colorado and to understand his thoughts when, fifty years later, old and spare, but straight, no longer a miner, but a philosopher, he lets his gaze float along with the smoke from his pipe over the ghostlike remnants of his town. The acceptance of New England pride, and the participation in the long suffering of the South.

And there is the switch of heroes.

Suppose that *you* go to live in a foreign country and that this country is Italy. And suppose you are talking to a cultivated Italian, who may say to you:

"Shakespeare? Pretty good, isn't he? There are Italian translations of Shakespeare, and some people read them. As for myself, I can read English and have read some of Lamb's *Tales from Shakespeare;* the dream in midsummer; Hamlet, the neurotic who could not make up his mind; and *Romeo* and *Giulietta*. Kind of queer ideas you Anglo-Saxons have about Italians! Anyhow, as I was saying, Shakespeare is pretty good. But all those historical figures he brings in . . . not the most important ones. . . . We have to look up history books to follow him.

"Now you take Dante. *Here* is a great poet for you! A universal poet! Such a superhuman conception of the universe! Such visions of the upper and nether worlds! The church is still walking in Dante's steps after more than six centuries. And his history! He has made history alive. Read Dante and you know history. . . ."

In your hero worship there is no place for both Shakespeare and Dante, and you must take your choice. If you are to live in Italy and be like other people, forget Shakespeare. Make a bonfire and sacrifice him, together with all American heroes, with Washington and Lincoln, Longfellow and Emerson, Bell and the Wright brothers. In the shadow of that cherry tree that Washington chopped down let an Italian warrior rest, and let him be a warrior with a blond beard and a red shirt. A warrior who on a white stallion, followed by a flamboyant handful of red-shirted youths, galloped and fought the length of the Italian peninsula to win it for a king, a warrior whose name is Garibaldi. Let Mazzini and Cavour replace Jefferson and Adams, Carducci and Manzoni take the place of Longfellow and Emerson. Learn that a population can be aroused not only by Paul Revere's night ride but also by the stone thrown by a little boy named Balilla. Forget that a telephone is a Bell telephone and accept Meucci as its inventor, and remember that the first idea of an airplane was Leonardo's. Once you have made these adjustments in your mind, you have become Italianized, perhaps. Perhaps you have not and never will.

When I travel across the immense plains of the Middle West, plowed and harvested at night by gremlins, for in the daytime no soul is ever to be seen, I still feel the impact of emptiness. I miss the crowded terraced fields reclaimed from the stony side of a hill. I miss the many eyes a tourist feels on his back—as an American friend once told me—each time he stops to eat his lunch in the most secluded spot in Italy. I miss the people who materialize from nowhere; the bashful peasant children with their hands behind their backs; the dark-haired girls who munch apples and throw inquisitive glances with their spicy eyes; the women who interrupt their chores, who wipe their hands on faded aprons and emerge from homes hidden out of sight by clumps of old trees; the men who were slumbering away the hot noon hours sprawled on the warm earth and now get up to view the tourist with the others.

If I still miss them, I ask myself, and still marvel at the vastness of America, at newly discovered sights, at the mention of some great name still unheard of by me, if I fail to understand the humor in Charles Addams' cartoons, can I truly say that I am Americanized?

(16)

SOME SHAPES OF THINGS TO COME

On January 16, 1939, two weeks after we had landed in America, Enrico and I went to the pier of the Swedish Line at an early hour of the afternoon. The liner "Drottningholm" was already in sight, inching its way toward us; and even before it came alongside the wharf, we recognized in a crowd the man we had come to meet, Professor Niels Bohr. He was standing by the rails of an upper deck, leaning forward, scanning the people on the dock.

We had seen Professor Bohr less than a month earlier, when, on our way from Stockholm to the United States, we had stopped in Copenhagen. We had spent most of our time in Professor Bohr's hospitable home, a beautiful villa, a little outside town, which a beer manufacturer had given to the most outstanding Dane for life use.

During the short time that had elapsed since our visit to his home, Professor Bohr seemed to have aged. For the last few months he had been extremely preoccupied about the political situation in Europe, and his worries showed on him. He stooped like a man carrying a heavy burden. His gaze, troubled and insecure, shifted from the one to the other of us, but stopped on none.

When he talked in the large noisy room at the pier, addressing nobody in particular, his soft, indistinct voice hardly reached my ears. His accent in English was different from that of any foreigner I knew. Of the words he spoke I grasped only the most familiar:

"Europe . . . war . . . Hitler . . . Denmark . . . danger . . . occupation. . . ."

From New York, Bohr proceeded to Princeton, where he had planned to spend a few months with Einstein. Princeton is not far from New York, and Bohr came often. I saw him several times and gradually became used to his way of speaking.

He talked of one subject only: the danger of war in Europe.

Despite the Munich agreement of September 29, 1938, Hitler had supported Hungary's and Poland's demands for parts of Czechoslovakia and allowed that country to be further dismembered.

Refugees arriving from all parts of Europe brought with them the gloom of impending disaster. Meanwhile, another dangerous character was coming into the limelight of the European scene: in December, 1938, Franco had undertaken his victorious campaign, which was to create one more totalitarian state.

Aware that the European system of security was collapsing, Professor Bohr worried for his family, his country, and all of Europe.

Two months after his landing in the United States, what remained of Czechoslovakia was annexed by Germany under the name of the "protectorate of Bohemia and Moravia"; and Bohr spoke about the doom of Europe in increasingly apocalyptic terms, and his face was that of a man haunted by one idea.

But to the physicists and to the other scientists who had occasion to approach him during the months of his stay in this country his mind seemed filled not so much with the darkening European social and political picture as with some recent scientific development: the discovery of uranium fission. In the light of subsequent events I must at least concede that in Bohr's mind there was room for both.

As I have discussed in an earlier chapter, uranium had been bombarded with neutrons, in the course of experiments performed in Rome in 1934, and a new element of atomic number 93 seemed to have been produced. A controversy over element 93 had started and dragged on inconclusively. Artificial radioactive elements were produced in such extremely small amounts that ordinary methods of separation and of chemical analysis could not be employed. A number of physicists and chemists had striven to develop special techniques, and great progress in this direction was achieved by a team of workers at the Kaiser Wilhelm Institute for Chemistry in Berlin: two chemists, Otto Hahn and Fritz Strassman, and a woman physicist, Lise Meitner. Although a Jew, she had been allowed to remain in Germany during the first years of the Nazi regime, because she was Austrian and her nationality made her immune from German anti-Semitic laws. After the *Anschluss*, however, she had been compelled to interrupt her work and leave Germany. Enrico and I met her in Stockholm in December, 1938, a worried, tired woman with the tense expression that all refugees had in common.

Hahn and Strassman carried on without Lise Meitner the work started with her, and toward the end of 1938 they were able to ascertain, by means of chemical processes, that some of the fragments obtained in bombarding uranium with slow neutrons were atoms of barium. Since an atom of barium is approximately half as heavy as an atom of uranium, it was logical to conclude that some uranium atoms had split into two almost equal parts. This kind of atomic disintegration had never been observed before. Atoms were known to break up and emit protons, which have mass 1; or neutrons, also of mass 1; occasionally an atomic fragment might have been an alpha particle, whose mass is 4. Never had an atom split into two big chunks, never had heavier fragments than alpha particles been detected. But the mass of the barium in these experiments was 139.

Hahn and Strassman succeeded in sending word of their results to Lise Meitner in Stockholm. At once she went to Copenhagen, where she and her nephew Otto Frisch, also a refugee from Germany, discussed the Hahn-Strassman experiment with Bohr, just prior to his departure for this country. They advanced the theory that when uranium split into two pieces, a process that Lise Meitner called *fission,* an enormous amount of nuclear energy must be released and that the two fragments must fly apart with tremendous speed. They also outlined an experiment to verify this hypothesis and to measure the amount of energy liberated in the fission of a uranium atom.

When Bohr arrived in America, a telegram was waiting for him: Meitner and Frisch had completed their experiment most successfully and had obtained results in agreement with their theory.

Enrico tried to explain fission to me. He does not like to talk shop at home, except when extraordinary events grant a breach in the rules, and the recent discovery well deserved the exceptional treatment. I am a slow thinker and lack the necessary background to grasp scientific matters easily.

"Wait a moment," I said, "let me see if I follow you. Hahn bombarded uranium atoms with slow neutrons . . ."

"That's right."

". . . and broke some of them into two big chunks . . ."

"Still correct."

". . . and uranium atoms would behave in the same way for anyone. . . ."

"What do you mean now?"

"I mean that *every time* uranium is bombarded with slow neutrons, at least part of it would break into two halves."

"That's right, again."

"But then . . . in Rome you and your friends also bombarded uranium with slow neutrons. Then if uranium behaves always in the same manner, you must have produced fission without recognizing it."

"That is exactly what happened. We did not have enough imagination to think that a different process of disintegration might occur in uranium than in any other element, and we tried to identify the radioactive products with elements close to uranium on the periodic table of elements. Moreover, we did not know enough chemistry to separate the products of uranium disintegration from one another, and we believed we had about four of them, while actually their number was closer to fifty."

"But then, what has become of your element 93?"

"What at the time we thought might be element 93 has proved to be a mixture of disintegration products. We had suspected it for a long time; now we are sure of it."

"Then," I thought to myself, "fission is the death warrant for your element 93."

To Enrico, however, the discovery of fission had deeper meaning than proving that an interpretation of experimental facts had been wrong. Foreseeing important developments, he set himself to work on the theory of the phenomenon.

Enrico has always oscillated between theoretical and experimental physics, conveniently adapting to changing needs. Whenever there seems to be no chance for an interesting experiment, Enrico withdraws to his office and fills sheet upon sheet with calculations. At home he appears absorbed, takes little notice of the family, and scribbles incomprehensible figures and notations on, the margins of newspapers: if I offer to bring him a pad of scratch paper, he declines on the grounds that he is doing nothing of importance. But as soon as he gets an idea for a piece of experimental research or whenever a new apparatus is being devised and completed, he lets his paper become covered with dust and spends all of his time in the laboratory.

When we came to this country he had left behind his Geiger counters, the Divine Providence's gram of radium, the substances gathered and stored to work upon. It was quite natural that at Columbia University he should go back to theoretical studies. During the preparation for leaving Italy, during the journey across Europe and the Atlantic, during the time it took to settle in New York, Enrico had seen no scientific literature. To regain the touch he had lost with research, he read a considerable number of papers. Enrico does not take long to do this. Many of his friends envy him his power of digesting hard scientific material at great speed. He reads papers only to the point where the statement of the problem is completed. Then he scribbles some calculations on a piece of paper, after which all he has to do is to compare the author's solution of the problem with his own.

When Enrico learned of fission, soon after Bohr's arrival, he was a theoretical physicist. He viewed fission from its theoretical aspects and soon advanced the hypothesis that when uranium splits in two it might emit neutrons.

To me neutrons are unimportant particles; they do not even carry an electric charge. It would seem of no importance whether new ones are freed from atoms or not. But as soon as Enrico formulated his hypothesis, many experimental physicists set themselves to find neutrons in fission with eagerness and manifest excitement. They knew what Enrico was talking about.

"It takes one neutron to split one atom of uranium," Enrico argued. "We must first produce and then use up that one neutron. Let's assume, however, that my hypothesis is correct and that an atom of uranium undergoing fission emits two neutrons. There would now be two neutrons available without the need of producing them. It is conceivable that they might hit two more atoms of uranium, split them, and make them emit two neutrons each. At the end of this second process of fission we would have four neutrons, which would split four atoms. After one more step, eight neutrons would be available and could split eight more atoms of uranium. In other words, starting with only a few man-produced neutrons to bombard a certain amount of uranium, we would be able to produce a set of reactions that would continue spontaneously until all uranium atoms were split."

This is the basic idea of a self-sustaining chain reaction.

The importance of a chain reaction lies in the fact that when an atom is violently torn into two pieces by a neutron, an enormous amount of energy is released, a fact which, as mentioned before, had already been verified by Meitner and Frisch. For the first time the possibilities of utilizing the unlimited sources of atomic energy appeared before human eyes. Atomic weapons were mentioned at a moment when war was probable. The thought that fission had been discovered in Germany was alarming: Would the Germans be able to run their battleships with atomic power or, even worse, to produce some kind of atomic explosion? There was still a hope: that to achieve a self-sustaining chain reaction would prove impossible in practice. The theoretical process of neutron emission outlined by Enrico was only ideal, for, in reality, not all neutrons emitted during fission would split more uranium atoms. Many would be absorbed by matter before they had a chance of hitting any uranium. Besides, neutrons produced in fission were too fast and not effective as atomic bullets, unless a method were found to slow them down.

The challenge of this complex problem spurred physicists to immediate action. Work in this field was undertaken in many universities, and at Columbia University the ball started rolling.

Shortly after his arrival in the United States Bohr went to Columbia in search of Enrico, but found Herbert Anderson instead.

Herbert must have shown no undue bashfulness, for Bohr talked to him at length of fission. Anderson listened with sharp attention and, as soon as Bohr left, rushed excitedly to look for Enrico.

"Why don't you plan some research in the field of fission with our cyclotron?" he asked. "I would like very much to work with you, and what better chance could there be than this?"

The experimentalist woke up in Enrico. He had never played with a cyclotron before.

A cyclotron is a machine that accelerates charged particles—for instance, protons. These particles, if not deviated by outside forces, travel in a straight line; in trying to accelerate them, one runs into the difficulty that they fly away too fast and are entirely out of reach before there is time to impart to them the desired speed. This difficulty was overcome by Ernest O. Lawrence when he devised his first cyclotron, which earned him the Nobel Prize: A big magnet bends the path of the particles and holds them inside a cylindrical box where they go round and round, faster and faster, until they have acquired a very high energy.

Particles emerging from the Columbia cyclotron, Anderson suggested, could be made to strike suitable substances and produce neutrons. The suggestion was the more attractive to Enrico in that it implied that he would go back to a kind of research that he had first started five years before. Still he was hesitant. Professor Pegram was the chairman of the physics department; John R. Dunning the man in direct charge of the cyclotron. It was up to them to organize the work.

"Still," Anderson insisted with remarkable determination, "I have built much equipment for the cyclotron. I have a right to work with it and to ask you to work also."

A compromise was finally reached between Anderson's enthusiasm and Enrico's scruples. At a conference called between Professor Pegram, Dunning, Fermi, and Anderson a plan of research was outlined. When Enrico emerged from this conference, he was again an experimental physicist. Besides, he had at his disposal a source of neutrons approximately a hundred thousand times more intense than the radon-beryllium sources in Rome: deuterons accelerated by the cyclotron were made to hit beryllium and could produce about 100,000 more neutrons per second than the sources in Rome at their best. Enrico's neutron sources were destined to jump by a factor of 100,000: the pile that he was going to use after the war was a source of neutrons that much more intense than the Columbia cyclotron.

Other physicists joined Enrico and Herbert, among them, the Hungarian-born Leo Szilard, and Walter H. Zinn, a Canadian-born, tall, blond young man who taught at City College and did research at Columbia.

For a while I could follow the progress of research at that distance which a layman finds himself from specialists. Once in a while Herbert Anderson or Wally Zinn or John Dunning would come to our house, and it might so happen that they talked shop with Enrico in front of me. I attended some lectures given by Enrico, I read a few reports in the newspapers. Soon, however, a voluntary system of censorship was established, and the lid of secrecy fell over nuclear physics. No word of it reached my ear for five whole years, from the summer of 1940 to that of 1945, when secrecy was partly lifted after an atomic bomb was dropped on Hiroshima.

I learned to ask no questions. No more "What have you done today?" nor "Are you pleased with your work?" nor "Who is your collaborator?" Enrico often went on mysterious trips. He would pack his suitcase and leave, instructing me to call his secretary if I needed to get in touch with him during his absence. On his return I was left alone with my speculations on the color of the mud under his shoes or on the amount of dust on his suit. Other women's husbands also went on frequent travels; it would have been bad taste indeed to ask where.

At about this time I received a present from a friend of mine, the wife of one of Enrico's colleagues. It was a novel by Harold Nicolson, entitled *Public Faces* and published in 1933, which related a diplomatic incident caused by the dropping of an atomic bomb.

(17)

AN ENEMY ALIEN WORKS FOR
UNCLE SAM

I found a copy of this letter more than ten years after it was written, when, in a fit of house-cleaning enthusiasm, I tackled a certain cabinet where we keep family papers.

March 16, 1939

Admiral S. C. Hooper
Office of Chief Naval Operations
Navy Department
Washington, D.C.

DEAR SIR:

. . . Experiments in the physics laboratories at Columbia University reveal that conditions may be found under which the chemical element uranium may be able to liberate its large excess of atomic energy, and that this might mean the possibility that uranium might be used as an explosive that would liberate a million times as much energy per pound as any known explosive. My own feeling is that the probabilities are against this, but my colleagues and I think that the bare possibility should not be disregarded, and I therefore telephoned . . . this morning chiefly to arrange a channel through which the results of our experiments might, if the occasion should arise, be transmitted to the proper authorities in the United States Navy.

Professor Enrico Fermi who, together with Dr. Szilard, Dr. Zinn, Mr. Anderson, and others, has been working on this problem in our laboratories, went to Washington this afternoon to lecture before the Philosophical Society in Washington this evening and will be in Washington tomorrow. He will telephone your office, and if you wish to see him will be glad to tell you more definitely what the state of the knowledge on this subject is at present.

Professor Fermi . . . is Professor of Physics at Columbia University

. . . was awarded the Nobel Prize. . . . There is no man more competent in this field of nuclear physics than Professor Fermi. . . .

Professor Fermi has recently arrived to stay permanently in this country and will become an American citizen in due course. . . .

Sincerely yours,

GEORGE B. PEGRAM

Professor of Physics

GBP:H

I had never set eyes on this letter and felt the excitement of the historian who discovers an important document. That same evening I showed it to Enrico.

At first he appeared to be as puzzled as I was. He read the letter over carefully, chewing on the end of his pencil in concentration. He turned to me.

"Where did you find it?" he asked.

"In a file labeled 'Miscellaneous,' together with a clipping from the newspaper *Lavoro Fascista* that criticized you because you did not give the Fascist salute to the King of. . . ."

"Now I remember!" Enrico interrupted. "I prepared that file when we became enemy aliens. I thought it might be used as evidence of our loyalty to the United States."

We became enemy aliens on December 8, 1941. On that day President Roosevelt declared that an "invasion or predatory incursion is threatened upon the territory of the United States" by Germany and Italy, and proclaimed Germans and Italians "alien enemies." Formal declaration of war to those countries occurred three days later.

I did not want this train of recollections to take me away from Professor Pegram's letter. I wanted to know the reasons why it was written.

It was self-explanatory, Enrico said. Professor Pegram had prepared it as an introduction to Admiral Hooper and had given a copy to Enrico so that he would be informed of its content.

"Did you really go see Admiral Hooper? What came out of your interview? Why did you never say anything about it?"

Enrico had seen the Admiral. To mention this fact would have been an indiscretion, although no official secrecy policy had yet been established. The interview had yielded little result.

163

"Couldn't you arouse the Admiral's interest in the atomic bomb?"

"You are using big words. You forget that in March, 1939, there was little likelihood of an atomic bomb, little proof that we were not pursuing a chimera." Enrico was no longer interested in this conversation. He had picked up the *New York Times*, and now he deployed it in front of him, thus indicating that the interview was at an end. I was left to my own thoughts.

Professor Pegram's letter, I reflected, is of historical significance as the first attempt of science to establish connections between research and government, not existing at that time. In this respect its most important part is its date: March 16. Only two months, to the day, had passed since Professor Niels Bohr had landed in America and received the celebrated telegram confirming uranium fission. During those two months experiments on fission had been performed at American universities, and Enrico's hypothesis that neutrons would be emitted had found experimental confirmation. The possibility of achieving a chain reaction and of placing the huge stores of energy in nuclei at man's disposal in a not too distant future had been envisaged and its implications probed. All this placed a burden of responsibility on the scientists too great for a small group to bear alone, even in a world at peace; and in March, 1939, the world was hardly at peace.

On that same March 16 when Professor Pegram had written his letter, Hitler had annexed what was left of Czechoslovakia after the Munich dismemberment. War was approaching. There could be little doubt of it. Results of nuclear studies ought not to be confined inside the laboratories. Hence the attempt at alerting the Navy.

It is not surprising that this attempt should have been inconclusive. Examined in the light of subsequent events, it appears to have been carried out with too great hesitation. That Fermi should wish to contact Admiral Hooper *because* he happened to be in Washington, that he should not plan his trip *in order* to see the Admiral, further minimizes that "bare possibility" of atomic explosives that sounds in itself overcasual, now that atomic weapons are a fact.

Professor Pegram's attitude was due to his cautious judgment that warned him against jumping to premature conclusions. His skepticism about the outcome of the work in his own laboratories was shared by many other scientists and was probably caused by a hope that nuclear weapons should prove unfeasible. And Enrico himself, when talking to Admiral Hooper, doubted the relevance of his predictions.

Skepticism and doubt, however, did not lighten the burden of responsibility on the physicists' shoulders. They tried to alert the government a few months later, and this time they succeeded.

Hungarian-born physicist Leo Szilard felt more strongly than anyone else in this matter of double responsibility, of the scientists toward the government and of the government toward that part of science that might become useful to the military.

Szilard talked repeatedly with his friends and aroused some of them. In July, 1939, he and another Hungarian-born physicist, Eugene Wigner, conferred with Einstein in Princeton. It is quite likely that they tried to evaluate the German progress in atomic research since Hahn's discovery of uranium fission. Several months had passed during which Teutonic efficiency could have achieved much. The American government should be informed of these matters with no delay.

The three men decided that they would prepare a letter to President Roosevelt and that Einstein would sign it, being by far the most prominent of all the scientists in the United States.

By the time the letter was ready, its content carefully planned and thoroughly discussed by several physicists, Einstein had gone for a rest at a remote place near Poconic on Long Island. A car was needed to reach him, and Szilard, who does not drive, engaged the help of a third Hungarian-born physicist, his young friend Edward Teller. It was August 2, 1939.

In his retreat Einstein received the physicists and read the letter.

SIR:

Some recent work by E. Fermi and L. Szilard, which has been communicated to me in manuscript, leads me to expect that the element uranium may be turned into a new and important souice of energy in the immediate future. . . .

Einstein's eyes slowly moved along the two full, typewritten pages. The letter called for watchfulness on the part of the Administration, explained in clear words the state of research, suggested a line of action. The letter also stated bluntly:

This new phenomenon would also lead to the construction of bombs, and it is conceivable—though much less certain—that extremely powerful bombs of a new type may thus be constructed.

Einstein reached the bottom of the second page:

"For the first time in history men will use energy that does not come from the sun," he commented and signed.

Szilard then asked economist Alexander Sachs to deliver Einstein's letter to the President. On October 11 Roosevelt received Sachs, read Einstein's missive, and listened to Sachs's explanations. At once the President appointed an "Advisory Committee on Uranium."

Why were the persons acting this drama all foreign-born? I asked myself. Why was Enrico chosen to go see Admiral Hooper, Enrico, a foreigner just arrived in this country, who still spoke with a thick accent and sprinkled whatever he said with a shower of extra vowels, who, in his friends' opinion, had never learned how to bang his fist on a table to get what he wanted?

In Italy, I reflected, no foreigner would have succeeded. . . . No, that line of thought brought me no closer to an answer: in Italy universities are government-controlled; a channel between universities and government is always in existence and does not need to be opened. That's it! These people, these Hungarian-, German-, and Italian-born, knew the organization in dictatorial countries; it *occurred* to them that there might be ties between research and military applications, that in Germany all scientific work might have been enrolled in the war effort. That is why President Roosevelt received his first warning from persons like Einstein, Szilard, Wigner, and Teller, while American-born and -raised physicists had not yet found the door out of their ivory tower: the first knew the military state and the concentration of powers, the latter had seen only democracy and free enterprise.

In the face of a national emergency totalitarian regimes are more efficient than democracies, at least during the first stages of the game. They are better equipped. A dictator holds in his hands all the strings and can pull them and mobilize the country at a moment's notice. A democracy has either no strings at all or long pieces of red tape. A dictator decrees; a president asks Congress for permission to organize.

After President Roosevelt appointed the Advisory Committee on Uranium, hearings, committee meetings, organization, and reorganization of boards, and shifts in directives kept research at no faster a pace than its natural one, and greatly limited government support. It was only on December 6, 1941, the day before Pearl Harbor, three years after the discovery of uranium fission in Germany, that the decision to make an all-out effort in atomic energy research was announced by Vannevar Bush, director of the Office of Scientific Research and Development.

Although limited Navy and Army contracts had been made prior to that date, it can be said with no great loss of accuracy that nuclear research became war work at the moment the United States entered the second World War. At that same moment Enrico found himself doing war work and became an enemy alien.

A number of our acquaintances were surprised to learn that we were enemy aliens.

"Why aren't you American citizens?" they asked. "Don't you want to become American citizens?"

American immigration and naturalization laws were little known before the war, and many ignored the basic fact that it takes five years of residence in the United States for an immigrant to become a citizen. More questions would have been asked, greater surprise displayed, had it been public knowledge that Enrico was engaged in war work.

I confess to some wonder myself. Did Uncle Sam take an unwarranted risk when he retained Enrico and a few other nationals of countries he was at war with in one of his most vital projects, that of the atomic bomb?

So far as Enrico is concerned the answer is contained in Professor Pegram's letter to Admiral Hooper. "There is no man more competent in this field of nuclear physics than Professor Fermi," Pegram had written in 1939. To dismiss Enrico because the project had become war work would not have wiped off his knowledge or his insight into the problems involved in construction of atomic bombs. And Uncle Sam would have lost the benefit of this knowledge and of this insight. So Enrico went on at his usual work, in his own field of research, which happened to have become of vital interest to Uncle Sam.

At the end of that same December, 1941, he took his first trip to Chicago. It was a cold winter in Chicago, and Enrico came home early in January with a fever and a touch of bronchitis. The trip had no other significance to me but to have been the cause of an illness. The words "chain reaction" and "atomic pile" were not in our home vocabulary.

Enrico spent the rest of the winter shuttling between Chicago and New York. His status of enemy alien required that every time he planned "to make a trip outside of his own community" he should "file a statement with the United States Attorney in his district at least 7 days prior to his departure." He was not allowed to leave unless he had received and carried along a copy of the "endorsement of the United States Attorney." (The quotes are from regulations governing travel of enemy aliens.)

Because we lived in New Jersey, Enrico was required to apply for his traveling permit to the state's attorney in Trenton. Making allowance for mailing and receiving letters, he had to plan his trips to Chicago some ten days in advance. Enrico never complained, never stressed the absurdity of the fact that while he traveled to do war work for the United States government, he was obliged to ask that same government for special travel permits. A law was a law, he used to say, and could not make distinctions between good and bad enemy aliens. Besides, it was wartime.

One morning, however, his permit had not come, and he was due to leave in the evening for Chicago. A telephone call to Trenton proved insufficient: the permit was granted and signed, but Enrico could not be exempted from carrying it with him. The secretary of the physics department at Columbia University was dispatched to Trenton and returned with Enrico's permit just in time for him to catch his train. He could not fly to Chicago because President Roosevelt had ruled that "no enemy alien shall undertake any air flight or ascend into the air."

Enrico strongly objects to people taking extra pains on his account, and the idea of the secretary's taking a train ride to Trenton in order to obtain his permit disturbed him greatly. This time he was thrown out of his imperturbability.

"If *they* want me to travel for them, *they'll* have to find a way to let me do so freely," he said.

Those mysterious *they*, who seemed to be responsible for all there had come to be unexplained in our life, managed to secure a permanent permit for Enrico to travel between New York and Chicago.

From the end of April, 1942, Enrico stayed permanently in Chicago, while I remained in Leonia with the children, to let them finish the school year there.

Enrico was unhappy to move. *They* (I did not know who they were) had decided to concentrate all *that* work (I did not know what it was) in Chicago and to enlarge it greatly, Enrico grumbled. It was the work he had started at Columbia with a small group of physicists. There is much to be said for a small group. It can work quite efficiently. Efficiency does not increase proportionally with number. A large group creates complicated administrative problems, and much effort is spent in organization.

There could be no better place than the United States in which to be nationals of an enemy country. Americans are a warmhearted and hospitable people. The fact that somewhere in their background they, too, were foreigners keeps them from discriminating against, and drives them to welcome, new immigrants. When war with Italy broke out, there were no reprisals against Italians and no personal ill feelings. Our neighbors were friendly and continued praising our "beautiful country."

I could not fail to remark how much this behavior of the Americans during the second World War differed from that of our own countrymen when Italy entered the first World War. There were violent demonstrations. Embassies and consulates were stoned. Almost all German and Austrian nationals fled Italy at once, fearing internment or arrest. A German man who used to live across the street from my parents' home stayed on for a week or two. I was a child then, but I remember my parents' remark that he would do well to leave Italy at once, for he was now an enemy. The word "enemy" was ominously leaned upon and acquired disproportionate connotations. They could not be reconciled with the old man who used to stop his slow paces when meeting us children on the hilly street and produce candy from his pockets. He would bend down and let his pince-nez dangle from its silk cord, while his smile set his white whiskers quivering like bushes under a gentle breeze. He did not leave his home after Italy had entered the war but was seen standing for hours by his window, a forlorn figure yearning for the lost friendliness of the street. The gossipy *portiere* of the small villas down the hill huddled together, and the word "spy" hovered in the air. Then he disappeared altogether.

We were much luckier than my German neighbor, and only minor annoyances were in store for us. We had lived under some strain since the beginning of the war in Europe. Hitler's easy and persistent victories during the first two years of the war had spread the belief that only the pessimists could be right in their predictions. There seemed to be little doubt that Germany would gain a final victory in Europe and that such a victory might mean a Nazi domination of America, not necessarily through actual invasion, but rather through increased power of the German Bunds, which would receive and follow precise directives from Hitler and his gang. If such an eventuality should come true, we would leave the United States. The position that Enrico had occupied in Italy and the work in which he was engaged here would make him an easy and conspicuous target for Nazi reprisals. It seemed only sensible to make plans.

Our friends the Mayers were as concerned as we were. We had first met them in Ann Arbor in 1930, when we had been on our first visit to America. They had then been newly wed; Joe a tall, blond, American boy; Maria a blond, medium-sized German girl from Göttingen, where they had met and married. Both were scientists, he a chemist, she a physicist. Because Joe had joined the faculty of Columbia University in the fall of 1939, they had bought a house in Leonia and had moved there at about the same time we did.

Maria, who still had many relatives in Germany, was well informed of what happened there and knew what naziism meant. The Mayers and the Fermis determined to leave the United States together if naziism should become established in this country. During the many evenings spent with the Mayers between the fall of France and America's entry into the war, we made plans together. Between a philological argument on the origin of some English word and a piece of advice on gardening that the Mayers passed down to the Fermis, we prepared to become modern Robinson Crusoes in some faraway desert island.

We made plans as soundly conceived in the theory, as carefully worked out in all details, as might be expected from a group which included two theoretical physicists and a practical, American-raised chemist.

Joe Mayer was to be our sea captain, a role in which he was not excessively experienced. Enrico's knowledge of currents, tides, and stars would help. His delight at the prospect of experimenting with compass and sextant was encouraging. Yet Joe felt we should practice navigation in the Florida waters at the first opportunity.

Meanwhile, there was much we could do. Maria Mayer and Enrico could consult and determine what part of our civilization was worth saving. Accordingly, Maria could collect the best-suited books. Enrico, the descendant of farmers, could study the agricultural problems of our refuge. It was my task to see that our colony would not go naked in years to come. I might decide on cotton seed and spinning wheels or on bolts of cloth. It did not matter, so long as everyone was clothed. A few scientifically selected persons would be invited to join our expedition: we ought to have a doctor; we ought to have children of such age, sex, and heredity that they could later marry ours and people our island.

What island we would make ours was still to be determined. In a war in which the United States would in all likelihood participate on the side against Germany, the Atlantic Ocean was out of the question. The Pacific Ocean is sown with islands. In the temperate zone between the Hawaiians and the Philippines there were numberless islets large enough for us. We would search for a desert island among them.

We could not foresee Pearl Harbor, and we overlooked the Japanese!

While envisaging adventure, Enrico and I did not neglect more practical precautions. Historical knowledge and personal experience had taught us that when war breaks out in a country, the assets of enemy aliens are immediately frozen. We could not predict the extent of American tolerance; we did not know that the financial restrictions would allow sufficient leeway for ample living. So we decided to bury a "treasure" in our basement. We were driven by that same need for action in the face of undefined but real danger that in 1936 had prompted us to procure gas masks for our family. In 1941 rather than feel powerless in the hands of Destiny, we prepared our financial survival: one evening we tiptoed down into the basement of our home after everybody else was in bed, like conspirators in the night. I swept the coal dust from the concrete floor of an old coalbin not in use, while Enrico held a flashlight for me. Then I held the flashlight, and he dug a hole in the concrete. He had placed a wad of notes in a lead pipe to protect them from the dampness of the ground, and we lowered pipe and "treasure" into the hole. I swept back some coal dust over the patch to conceal it, and we went to bed with a lighter conscience than on the nights before.

The "treasure" proved unnecessary. We dug it up only when we left Leonia.

When the United States entered the war and we became enemy aliens, we did other things which subsequently proved meaningless and unnecessary but gave us the illusion that we were in control of events. Enrico prepared his *miscellaneous* file for the benefit of FBI men who might have questioned our loyalty but never did. We burned Nella's second-grade reader, which had come from Italy with our furniture and our books. It had seemed an incriminating piece of property, because it contained so many pictures of Mussolini: Mussolini as a plain Fascist in a black shirt; Mussolini in black fez, shiny black boots, and full uniform of black *orbace,* a material especially woven for Fascist uniforms; Mussolini riding on a beautiful horse; Mussolini paternally smiling upon little "sons and daughters of the she-wolf"; Mussolini inspecting parades of boys from eight years of age up, in military apparel, rifles across their frail shoulders. Altogether too many Mussolinis!

No matter how many precautions one may take against the un foreseen, one invariably overlooks a detail which later becomes the source of concern. The overlooked detail in our case was our five-year-old son Giulio. Giulio, some friends told us, was going around Leonia saying that he wished Hitler and Mussolini would win the war. Friends of ours had overheard him. And what is the origin of children's opinions, if not the home?

In peacetime nobody could have taken Giulio seriously. But at the beginning of the war minds were filled with an insecurity that was no less pervasive for being still vague and not related to specific facts. It distorted judgments, made people suspicious—even the quiet people of Leonia. When a stranger with a German accent had come to see a house for sale in Leonia, many eyes had followed him. When he had climbed on the roof to inspect it with an accuracy that seemed excessive, rumors had started: was he going to receive signals from the air?

Thus Giulio's words might be taken seriously, we feared, despite the fact that he had also joined other little boys in singing:

> We'll wipe the Japs
> Out of the maps,

and that the two contradictory attitudes should balance one another.

We summoned Giulio and asked him why he had expressed such silly ideas about Hitler and Mussolini. Of course, he did not know. But to me the explanation was evident. Giulio had often been reminded that he, of all the members of our family, had seen the two dictators together, when Hitler had come to Rome for a visit. He had witnessed a historical event that few had witnessed. In his mind, perhaps, he had come to consider the two dictators as his private property. Besides, because he was small for his age and delicate, Giulio always felt the need of bragging, of telling tall tales, of somehow making himself conspicuous.

Under our insistent questioning Giulio broke into sobs:

"I was joking; I didn't mean it!"

Enrico had no mercy. Giulio's nonsense must be stopped.

"Suppose a responsible citizen reports you. Suppose an FBI man overhears you. What do you think they would do? Wouldn't it be their duty to put you in jail?"

Giulio cried some more, and then dried the tears from his large brown eyes, blew his nose, and went to play, as any other small boy would have done.

The children and I stayed in Leonia until the end of June, 1942. As soon as school was over, I shipped Nella and Giulio to a camp in New England, and I joined Enrico in Chicago.

I had come to consider Leonia as our permanent home and loathed the idea of moving again.

Enrico thought we would be in Chicago for the duration and then we would go back to Leonia. He was an optimist. We had not learned yet the definition of *duration* accepted in some (limited) circles: the time it would take for all physicists on the East Coast to reach the West Coast and for all physicists on the West Coast to reach the East Coast. I did not need this joke to feel the uncertainty of the future. Before leaving Leonia I went to say goodbye to the Ureys. With his kind eyes full of concern, Harold said to me:

"Laura, you'll never come back to Leonia."

That gave me material to ponder upon, and it was then that I made my resolution never to grow roots again. Harold Urey was only half a prophet, for he had not foreseen that he, too, would leave Leonia and go to Chicago. But that was three years later.

Meanwhile, I looked for a house in Chicago and found one that seemed to fill our requirements. It was pleasant, well furnished, and conveniently located near the University of Chicago campus. The owner, a businessman, was going to Washington with his family "for the duration," for exactly the same period of time we would be in Chicago. There were, however, two sets of difficulties. the beautiful Capehart radio in the living room was equipped with a short-wave set and on the third floor there lived two Japanese-American girls who would have liked to stay. Regulations on enemy aliens forbade us to own or use a short-wave set, and, Enrico said, an Italian family plus two Japanese girls made a nest of spies.

There was clearly no housing shortage in Chicago, or our landlord would have looked for easier tenants. He checked with the FBI instead and was informed that, despite the fact that Enrico was doing war work, we could not be in possession of a radio with short waves. The landlord requested the Capehart factory to block off the incriminating waves; and he asked the two Japanese girls to move.

On October 12 of that same year, Attorney General Biddle, on the occasion of the Columbus Day celebration, announced that Italians would no longer be considered enemy aliens. Now we could travel without permits, have the short waves reinstalled in the radio, and be in possession of cameras and binoculars.

On July 11, 1944, Enrico and I swore allegiance to the United States of America in the District Court of the United States at Chicago. We were at last American citizens after five and a half years of residence in this country.

OF SECRECY AND THE PILE

The period of great secrecy in our life started when we moved to Chicago. Enrico walked to work every morning. Not to the physics building, nor simply to the "lab," but to the "Met. Lab.," the Metallurgical Laboratory. Everything was top secret there. I was told one single secret: there were no metallurgists at the Metallurgical Laboratory. Even this piece of information was not to be divulged. As a matter of fact, the less I talked, the better; the fewer people I saw outside the group working at the Met. Lab., the wiser I would be.

In the fall Mr. and Mrs. Arthur H. Compton—I was to learn later that he was in charge of the Metallurgical Project—gave a series of parties for newcomers at the Metallurgical Laboratory. Newcomers were by then so numerous that not even in Ida Noyes Hall, the students' recreation hall, was there a room large enough to seat them all at once; so they were invited in shifts. At each of these parties the English film *Next of Kin* was shown. It depicted in dark tones the consequences of negligence and carelessness. A briefcase laid down on the floor in a public place is stolen by a spy. English military plans become known to the enemy. Bombardments, destruction of civilian homes, and an unnecessary high toll of lives on the fighting front are the result.

After the film there was no need for words.

Willingly we accepted the hint and confined our social activities to the group of "metallurgists." Its always expanding size provided ample possibilities of choice; besides, most of them were congenial, as was to be expected, for they were scientists.

The nonworking wives wished, quite understandably, to do something for the war effort. One of the possible activities along this line was to help entertain the armed forces at the USO. I preferred to sew for the Red Cross or to work as a volunteer in the hospital of the university, and to save my social capacities for the people at the Met. Lab., who had not the benefit of the USO.

Thus early in December, 1942, I gave a large party for the metallurgists who worked with Enrico and for their wives. As the first bell rang shortly after eight in the evening, Enrico went to open the door, and I kept a few steps behind him in the hall. Walter Zinn and his wife Jean walked in, bringing along the icy-cold air that clung to their clothes. Their teeth chattered. They shook the snow from their shoulders and stamped their feet heavily on the floor to reactivate the circulation in limbs made numb by the subzero weather. Walter extended his hand to Enrico and said:

"Congratulations."

"Congratulations?" I asked, puzzled, "What for?" Nobody took any notice of me.

Enrico was busy hanging Jean's coat in the closet, and both the Zinns were fumbling at their snow boots with sluggish fingers.

"Nasty weather," Jean said, getting up from her bent position to put her boots in a corner. Walter again stamped his feet noisily on the floor.

"Won't you come into the living-room?" Enrico asked. Before we had time to sit down, the bell rang again; again Enrico went to open the door, and amid repeated stamping of feet and complaints about the extraordinarily cold weather I again heard a man's voice:

"Congratulations."

It went on this same way until all our guests had arrived. Every single man congratulated Enrico. He accepted the congratulations readily, with no embarrassment or show of modesty, with no words, but with a steady grin on his face.

My inquiries received either no answer at all or such evasive replies as: "Ask your husband," or: "Nothing special. He is a smart guy. That's all," or: "Don't get excited. You'll find out sometime."

I had nothing to help me guess. Enrico had mentioned nothing worthy of notice, and nothing unusual had happened, except, of course, the preparations for the party. And those did not involve Enrico and provided no ground for congratulating.

I had cleaned house all morning; I had polished silver. I had picked up the electric train in Giulio's room and the books in Nella's. If there is a formula to teach order to children, I have not found it. I had run the vacuum, dusted, and sighed. All along I was making calculations in my mind:

"Half an hour to set the table. Half an hour to spread sandwiches. Half an hour to collect juices for the punch. . . . I must remember to make tea for my punch soon, so that it will have time to cool. . . . And if people start coming by eight, we'll have to start dressing by seven-thirty, and eating dinner by. . . ." So I had calculated my afternoon schedule backward from the time the company would arrive up to when I should set myself to work.

My schedule was upset, as schedules will be. While I was baking cookies in the kitchen, the house had gone surprisingly quiet, too quiet to contain Giulio and his two girl friends who had come to play. Where were they? Into what sort of mischief had they got themselves? I found them on the third-floor porch. The three angelic-looking little children were mixing snow with the soil in the flower pots and throwing balls at our neighbor's recently washed windows. Precious time was spent in scolding and punishing, in seeing what could be done to placate our neighbor.

So at dinner time Enrico found me hurrying through the last preparations, absorbed in my task and even less than usually inclined to ask questions of him. We rushed through dinner, and then I realized we had no cigarettes. It was not unusual: we don't smoke, and I always forget to buy them.

"Enrico, wouldn't you run to the drugstore for cigarettes?" I asked. The answer was what I expected, what it had been on other such occasions:

"I don't know how to buy them."

"We can't do without cigarettes for our guests," I insisted, as I always did; "it isn't done."

"We'll set the habit, then. Besides, the less our company smokes, the better. Not so much foul smell in the house tomorrow."

This little act was almost a ritual performed before each party. There was nothing unusual in it, nor in Enrico's behavior. Then why the congratulations?

I went up to Leona Woods, a tall young girl built like an athlete, who could do a man's job and do it well. She was the only woman physicist in Enrico's group. At that time her mother, who was also endowed with inexhaustible energy, was running a small farm near Chicago almost by herself. To relieve Mrs. Woods of some work, Leona divided her time and her allegiance between atoms and potatoes. Because I refused either to smash atoms or to dig potatoes, she looked down on me. I had been at the Woods's farm, however, and had helped with picking apples. Leona, I thought, owed me some friendliness.

"Leona, be kind. Tell me what Enrico did to earn these congratulations."

Leona bent her head, covered with short, deep-black hair, toward me, and from her lips came a whisper:

"He has sunk a Japanese admiral."

"You are making fun of me," I protested.

But Herbert Anderson came to join forces with Leona. Herbert, the boy who had been a graduate student at Columbia University when we arrived in the United States, had taken his Ph.D. with Enrico and was still working with him. He had come to Chicago a few months before I did.

"Do you think anything is impossible for Enrico?" he asked me with an earnest, almost chiding, face.

No matter how firmly the logical part of my mind did disbelieve, there still was another, way back, almost in the subconscious, that was fighting for acceptance of Leona's and Herbert's words. Herbert was Enrico's mentor. Leona, who was young enough to have submitted to intelligence tests in her recent school days, was said to have a spectacular I.Q. They should know. To sink a ship in the Pacific from Chicago . . . perhaps power rays were discovered. . . .

When a struggle between two parts of one's mind is not promptly resolved with clear outcome, doubt results. My doubt was to last a long time.

That evening no more was said about admirals. The party proceeded as most parties do, with a great deal of small talk around the punch bowl in the dining-room; with comments on the war in the living-room; with games of ping-pong and shuffleboard on the third floor, because Enrico has always enjoyed playing games, and most of our guests were young.

In the days that followed I made vain efforts to clear my doubts.

"Enrico, did you really sink a Japanese admiral?"

"Did I?" Enrico would answer with a candid expression.

"So you did not sink a Japanese admiral!"

"Didn't I?" His expression would not change.

Two years and a half elapsed. One evening, shortly after the end of the war in Japan, Enrico brought home a mimeographed, paper-bound volume.

"It may interest you to see the Smyth Report," he said. "It contains all declassified information on atomic energy. It was just released for publication, and this is an advance copy."

It was not easy reading. I struggled with its technical language and its difficult content until slowly, painfully, I worked my way through it. When I reached the middle of the book, I found the reason for the congratulations Enrico had received at our party. On the afternoon of that day, December 2, 1942, the first chain reaction was achieved and the first atomic pile operated success-fully, under Enrico's direction. Young Leona Woods had considered this feat equivalent to the sinking of an admiral's ship with the admiral inside. The atomic bomb still lay in the womb of the future, and Leona could not foresee Hiroshima.

The operation of the atomic pile was the result of almost four years of sustained work, which started when discovery of uranium fission became known, arousing enormous interest among physicists.

Experiments at Columbia University, I have already explained, and at other universities in the United States had confirmed Enrico's hypothesis that neutrons would be emitted in the process of fission. Consequently, a chain reaction appeared possible in theory. To achieve it in practice seemed a vague and distant possibility. The odds against it were so great that only the small group of stubborn physicists at Columbia pursued work in that direction. At once they were faced with two sets of difficulties.

The first lay in the fact that neutrons emitted in the process of uranium fission were too fast to be effective atomic bullets and to cause fission in uranium. The second difficulty was due to loss of neutrons: under normal circumstances most of the neutrons produced in fission escaped into the air or were absorbed by matter before they had a chance of acting as uranium splitters. Too few produced fission to cause a chain reaction.

Neutrons would have to be slowed down and their losses reduced by a large factor, if a chain reaction was to be achieved. Was this feasible?

To slow down neutrons was an old trick for Enrico, from the time when he and his friends in Rome had recognized the extraordinary action of paraffin and water on neutrons. So the group at Columbia—Szilard, Zinn, Anderson, and Enrico—undertook the investigation of fission of uranium under water. Water, in the physicists' language, was being used as a moderator.

After many months of research they came to the conclusion that neither water nor any other hydrogenated substance is a suitable moderator. Hydrogen absorbs too many neutrons and makes a chain reaction impossible.

Leo Szilard and Fermi suggested trying carbon for a moderator. They thought that carbon would slow down neutrons sufficiently and absorb fewer of them than water, provided it was of a high degree of purity. Impurities have an astounding capacity for swallowing neutrons.

Szilard and Fermi conceived a contrivance that they thought might produce a chain reaction. It would be made of uranium and very pure graphite disposed in layers: layers exclusively of graphite would alternate with layers in which uranium chunks would be imbedded in graphite. In other words, it would be a "pile."

An atomic pile is, of necessity, a bulky object. If it were too small, neutrons would escape into the surrounding air before they had a chance to hit a uranium atom, and they would be lost to fission and chain reaction. How large the pile ought to be, nobody knew.

Did it matter whether the scientists did not know the size of the pile? All they had to do, one might think, was to put blocks of graphite over blocks of graphite, alternating them with lumps of uranium, and keep on at it until they had reached the critical size, at which a chain reaction would occur. They could also give the pile different shapes—cubical, pyramidal, oval, spherical—and determine which worked best.

It was not so simple. Only a few grams of metallic uranium were available in the United States, and no commercial graphite came close to the requirements of purity.

The 1951 edition of Webster's *New Collegiate Dictionary* states that graphite is "soft, black, native carbon of metallic luster; often called *plumbago* or *black lead*. It is used for lead pencils, crucibles, lubricants, etc. . . ." The atomic pile built in 1942, clearly included in the "etc.," was to use as much graphite as would go into making a pencil for each inhabitant of the earth, man, woman, and child. Moreover, graphite for a pile must be of a state of purity absolutely inconceivable for any other purpose. Scientists would have to be patient.

Procurement became a big and important task, one for which Fermi was not suited and which he would rather leave to others. Luckily for him, Leo Szilard did not share his aversion to interrupting research and shopping around.

Szilard was a man with an astounding number of ideas, several of which turned out to be good. He had no fewer acquaintances than ideas, a not negligible percentage of whom were important persons in high positions. These two sets of circumstances made of Szilard a powerful and useful spokesman for the small group of researchers, one who could confront the difficulties of politics with sufficient impetus to overcome them successfully. Willingly and with determination he undertook the not easy task of turning grams into tons, both of metallic uranium and of highly pure graphite.

The first question one asks when undertaking a task of that kind is: "Who is going to finance my enterprise and give me the cash that's needed?" Szilard hoped he knew the answer. During the summer of 1939, with Wigner, Teller, Einstein, and Sachs, he had succeeded in arousing President Roosevelt's interest in uranium work. Now, at the very beginning of 1940, he scored his second victory and obtained the first tangible, if small, proof of that professed interest, when Columbia University received the first grant of $6,000 from the Army and Navy to purchase materials.

Thus by early spring 1940 a few tons of pure graphite started to arrive at the physics building of Columbia University. Fermi and Anderson turned into bricklayers and began to stack graphite bricks in one of their laboratories.

They were well aware that for many months, perhaps for years, there would not be uranium and graphite of good enough quality and in sufficient quantity to attempt a pile. That did not matter for the time being: they knew so very little about the properties of the substances they were to work with—of metallic uranium not even the melting point had been determined—that much study of these properties ought to be pursued and completed before they could in good conscience recommend that the Uranium Committee undertake the tremendous effort and expense that would go in the project.

So they stacked graphite bricks into a stocky column, placed a neutron source under it, observed what happened to the neutrons in the graphite, and began to collect data.

This work, dull as it sounds, was considered very important; and when the Advisory Committee on Uranium met on April 28, 1940, it decided to wait for more results at Columbia University before making formal recommendations for the project. The committee made this decision despite the report that the Nazis had set aside a large section of the Kaiser Wilhelm Institute in Berlin for research on uranium.

After the study on graphite, came that on uranium: How does it absorb and re-emit neutrons? Under what conditions will it undergo fission? How many neutrons will be produced altogether?

The experiments proceeded slowly for lack of materials, and Fermi would have liked to speed up his work. Besides, he was convinced that from the behavior of a small pile he would obtain much information pertinent to building a larger pile. Fermi and his group were able to start work on the "small pile" by the spring of 1941. They demolished their column of graphite bricks and laid them down again, placing lumps of uranium among them. Slowly, as more graphite arrived at Columbia, a black wall grew up. The black wall reached the ceiling; but it was still far from being a chain-reacting pile: too many neutrons escaped from it or were absorbed inside it, and too few remained to produce fission.

It became evident that the experiment could not be pursued to final success in that same laboratory. A larger room, with higher ceiling, was needed. No such room was available at Columbia, and somebody would have to look for one elsewhere. Fermi was absorbed in his research. His work was too important to be interrupted. So Herbert Anderson took off his overalls, put on a suit coat and a hat, and went scouting New York City and its suburbs in search of a loft that could house a pile. He spotted several possibilities and began some bargaining aimed at the best deal.

Before Herbert could take a final choice, Enrico learned that he, his group, his equipment, and the materials he had gathered would have to move to Chicago. It was the very end of 1941.

A few days previously, on December 6, Vannevar Bush, who was at the head of the Uranium Project, had announced that an all-out effort would be made to speed up atomic research. He had then assigned the various responsibilities to different top men and had placed Professor Arthur H. Compton, of the University of Chicago, in charge of fundamental physical studies of the chain reaction. Soon afterward Compton had resolved to bring all work under him to Chicago.

After the all-out effort was announced, the Uranium Project spread out like a huge octopus whose tentacles reached the most important universities and industries of the United States. The octopus had had its first spurt of growing early in 1941, when interest in the feasibility of a chain reaction had expanded from Columbia to a few other universities. But the baby octopus shot suddenly into adolescence, without going through childhood, in early 1942.

By that time the Uranium Project included more than nuclear research. There was production of graphite, of uranium, both as metal and as oxide; there was separation of uranium isotopes, some of which would undergo fission more readily than others; and there was production of a new element, plutonium.

The discovery of plutonium 239, the important isotope of plutonium, is due to a group of chemists and physicists at the University of California in Berkeley. Our old friend Emilio Segrè was in that group.

Emilio, whom we had not seen since he had left Italy in the summer of 1938, came to visit us in Leonia during Christmas vacation in 1940. He and Enrico took long walks together on the Palisades along the Hudson River, where they are covered with trees above the bare rock that falls precipitously into the water. They exchanged family news and the information each had of old friends. They made gloomy remarks about the war—the Germans had not yet invaded England, as many had predicted they would after the fall of France; but would they still?—remarks that made Emilio's lower lip come forward in concern. But they also talked of a substance that had not been discovered. Many physicists knew from theoretical considerations that element 94 might be found among the products of uranium reactions. It seemed probable that it would be as fissionable as the most fissionable of uranium isotopes. Enrico and Emilio discussed its likely chemical and physical properties. They concluded that it would be a good thing if researchers in California were to use their powerful cyclotron to produce this element and would direct their efforts to separating and identifying it.

Emilio went back to Berkeley, and soon the group to which he belonged discovered plutonium 239.

When Professor Compton was placed in charge of nuclear research, Enrico took his first trip to Chicago and applied for his first travel permit as an enemy alien. So the mysterious "they" for whom he traveled were none other than Compton and the policy-makers at the Uranium Project.

Had I been aware that the decision of moving work from Columbia to Chicago was due to Compton, I might not have grumbled. Compton was a thoughtful and considerate person, who took no step without weighing its effects on others. Perhaps because of this, whenever he expressed an opinion, it was interpreted as an order and accepted without much comment.

Occasionally, though, even Compton had to resort to expedients. Once he made a decision he could not explain because of secrecy. Compton called a meeting of all "metallurgists." They gathered in a hall, and Compton walked in, his deep-set eyes lowered on a Bible that he carried in his hands. He opened it. From it he read aloud:

"And God said unto Gideon . . ."

As Gideon accepted God's command, so the "metallurgists" should accept Compton's decision.

When Compton resolved to concentrate all work on the chain reaction in Chicago as soon as possible, Enrico's group was split. Anderson moved to Chicago at once, Zinn stayed at Columbia for a few months. Enrico shuttled between the two and applied for more traveling permits.

Zinn went on studying the pile built at Columbia, while Anderson started one of his own in Chicago. The Columbia pile had already reached the ceiling and could go no higher. Enrico wished to determine how much he could improve its performance by removing the air from it. Graphite is a porous substance and much air is trapped in its tiny pores. This air contributes to neutron absorption. If a vacuum were to be made in the space around the pile, the graphite pores would also become devoid of air, and a cause of neutron losses would be eliminated.

It is easy to say "Let's pump the air off the pile." To do it is a different matter.

Pumping air out of the entire room in which the pile was built would be impractical, even if feasible, for no one could work in a vacuum, at least without wearing clumsy diving suits.

"How is air removed in practice?" Enrico asked himself. Whenever air is to be taken away from food, the food is inclosed in a tin can. Why not do the same to the pile? Why not *can* the pile? There are no ready-made cans of the needed size, so Enrico ordered one. Tinsmiths built the huge can in sections. To insure proper assembly, they marked each section with a little figure of a man: if the can were put together as it should be, all men would stand on their feet, otherwise on their heads.

Once the pile was canned, vacuum pumps removed the air from it. The effect on neutron loss was appreciable, but not great. Perhaps, Enrico thought, pumping into the can methane, a nonabsorbing gas, which would fill the graphite pores, might further improve the pile performance. But when methane mixes with air, it may explode, and Fermi did not want to take more risks than those already being incurred. Handling radioactive substances, being exposed to larger numbers of neutrons than ever in the past, working with substances whose effects on human beings had not been tested—these were all hazardous conditions.

There had also been accidents. Walter Zinn had opened a can of thorium powder. It was known that powdered metals occasionally burned when coming in contact with air, and Zinn had taken sensible precautions. He had put on goggles, rubber gloves, and a rubber apron. As he opened his can of thorium, it exploded in his hands. His gloves caught on fire. His hands and face were severely burned. The goggles saved his eyes. He was hospitalized for several weeks. Even as he was discussing with Fermi the experiment with methane, his face and hands were reddened and scarred.

Another accident seemed to have had no ill consequences. For some experiments it was convenient to work with neutrons irradiating from a small concentrated source, as from a point. In that case the cyclotron was not satisfactory because it was too diffuse a source of neutrons. Then sources similar to those used in Rome were prepared at Columbia.

Enrico and his friends in Rome had mixed beryllium powder with the radon that had formed from the one gram of radium of Professor Trabacchi, the "Divine Providence." Radon decayed relatively fast, and each week they had extracted more and prepared new sources. It would have been more convenient to use the radium directly, but the "Divine Providence" had only that one gram of radium and could not dispose of it because it belonged to the Sanità Publica, the Italian Health Department. Consequently, in Rome the physicists had to be satisfied with using the product of radium, not radium itself.

Uncle Sam is wealthier than the "Divine Providence." He gave the researchers two grams of radium for keeps, to be used as they deemed best. Pegram, Fermi, and Anderson ground radium and beryllium for their sources and reduced them to powder. One day the powder was not sufficiently dry for their purposes, and they placed it on a hot plate to dry. They knew that radium would contaminate the air in the room where it was being heated, so they stayed outside it. Now and then they opened the door a little, just enough to peek in, and watched their cooking.

Unaccountably their mixture burned. When they opened their usual crack in the door they saw the room full of fumes. They rushed inside to turn off the plate and ran out immediately. They had been exposed to the fumes only a few seconds. A check with Geiger counters indicated that no radium had deposited in their bodies, and they set their minds at ease.

Five or six years went by, then Herbert became ill. Doctors diagnosed his illness as berylliosis, a very rare disease caused by deposit of beryllium in the lungs. Nobody could foresee Herbert's illness, because at the time he breathed the fumes of burned beryllium it was not known that beryllium had harmful effects. Nevertheless, Enrico did not relish the idea of increasing the risks of his work by handling the large quantity of methane needed to fill the canned pile. In the end he decided against it.

Work is only part of life. Men's emotions mingle and cannot be kept separate and confined to certain hours. When Fermi made his decision, a student let go a sigh of relief that had nothing to do with fear of being blasted by a methane explosion.

This student was Harold Agnew. He had come to New York from Chicago, where he worked with Herbert Anderson. Zinn's group at Columbia was to move to the Met. Lab. in Chicago as soon as the experiments under way were completed. Anderson, anxious to be reunited with Zinn, sent Agnew and two other young people to New York to help speed up the experiments and pack the equipment. They would not have to stay long in New York, Herbert said. There was no reason why Harold Agnew should change his plans. He would be back in time to get married in his native Denver on the date he had set. When Harold learned that a new experiment was under consideration, he feared a delay of his wedding. But now he was happy.

Harold Agnew had joined Anderson's group at the Met. Lab. in early February of that year. Agnew had completed his requirements for a B.A. at the University of Denver, had obtained a fellowship for graduate work at various universities, and his only embarrassment was having to make a choice. He sought advice from his teacher, Joyce Stearns.

Professor Stearns was going to join the Metallurgical Project in Chicago, he told Agnew. His friend Arthur Compton had persuaded him to go work there. If he wanted, Agnew could go along with him. The best men in physics, Stearns pointed out to Agnew, had left the universities and were now at various war projects. He would learn more physics at the Met. Lab. than at any of the universities.

The young man could do little but take his teacher's word: security regulations prevented Stearns from mentioning the men who would be in Chicago or from explaining what sort of work the boy would do there.

Then and in the next few months other brilliant students joined the project for the same reasons: at the Met. Lab., besides doing their part to save our country, they would learn much more physics than at depleted universities.

In Chicago, Harold Agnew was assigned to Anderson's group, and in the spring Anderson asked him to go help Fermi in New York. Once there was no more danger of postponing his wedding, Harold participated in the work of dismantling and packing the pile with renewed alacrity and zeal.

Graphite and uranium, radium and beryllium sources, Geiger counters and other instruments—everything was packed and shipped to the Metallurgical Laboratory. And Enrico went to settle in Chicago.

SUCCESS

Meanwhile Herbert Anderson and his group at the Met. Lab. had also been building small piles and gathering information for a larger pile from their behavior. The best place Compton had been able to find for work on the pile was a squash court under the West Stands of Stagg Field, the University of Chicago stadium. President Hutchins had banned football from the Chicago campus, and Stagg Field was used for odd purposes. To the west, on Ellis Avenue, the stadium is closed by a tall gray-stone structure in the guise of a medieval castle. Through a heavy portal is the entrance to the space beneath the West Stands. The Squash Court was part of this space. It was 30 feet wide, twice as long, and over 26 feet high.

The physicists would have liked more space, but places better suited for the pile, which Professor Compton had hoped he could have, had been requisitioned by the expanding armed forces stationed in Chicago. The physicists were to be contented with the Squash Court, and there Herbert Anderson had started assembling piles. They were still "small piles," because material flowed to the West Stands at a very slow, if steady, pace. As each new shipment of crates arrived, Herbert's spirits rose. He loved working and was of impatient temperament. His slender, almost delicate, body had unsuspected resilience and endurance. He could work at all hours and drive his associates to work along with his same intensity and enthusiasm.

A shipment of crates arrived at the West Stands on a Saturday afternoon, when the hired men who would normally unpack them were not working. A university professor, older by several years than Herbert, gave a look at the crates and said lightly: "Those fellows will unpack them Monday morning."

"Those fellows, Hell! We'll do them now," flared up Herbert, who had never felt inhibited in the presence of older men, higher up in the academic hierarchy. The professor took off his coat, and the two of them started wrenching at the crates.

Profanity was freely used at the Met. Lab. It relieved the tension built up by having to work against time. Would Germany get atomic weapons before the United States developed them? Would these weapons come in time to help win the war? These unanswered questions constantly present in the minds of the leaders in the project pressed them to work faster and faster, to be tense, and to swear.

Success was assured by the spring. A small pile assembled in the Squash Court showed that all conditions—purity of materials, distribution of uranium in the graphite lattice—were such that a pile of critical size would chain-react.

"It could be May, or early June at latest," Enrico told me, as we recently reminisced about the times of the Met. Lab. "I remember I talked about that experiment on the Indiana dunes, and it was the first time I saw the dunes. You were still in Leonia. I went with a group from the Met. Lab. I liked the dunes: it was a clear day, with no fog to dim colors. . . ."

"I don't want to hear about the dunes," I said. "Tell me about that experiment."

"I like to swim in the lake, . . ." Enrico paid no attention to my remark. I knew that he enjoyed a good swim, and I could well imagine him challenging a group of younger people, swimming farther and for a longer time than any of them, then emerging on the shore with a triumphant grin.

"Tell me about that experiment," I insisted.

"We came out of the water, and we walked along the beach."

I began to feel impatient. He did not have to mention the walk. He always walks after swimming, dripping wet, water streaming from his hair. In 1942 there was certainly much more hair on his head to shed water, not just the little fringe on the sides and on the back that there is now, and it was much darker.

". . . and I talked about the experiment with Professor Stearns. The two of us walked ahead of the others on the beach. I remember our efforts to speak in such a way that the others would not understand. . . ."

"Why? Didn't everyone at the Met. Lab. know that you were building piles?"

"They knew we built piles. They did not know that at last we had the certainty that a pile would work. The fact that a chain reaction was feasible remained classified material for a while. I could talk freely with Stearns because he was one of the leaders."

"If you were sure a larger pile would work, why didn't you start it at once?"

"We did not have enough materials, neither uranium nor graphite. Procurement of uranium metal was always an obstacle. It hampered progress."

While waiting for more materials, Herbert Anderson went to the Goodyear Tire and Rubber Company to place an order for a square balloon. The Goodyear people had never heard of square balloons, they did not think they could fly. At first they threw suspicious glances at Herbert. The young man, however, seemed to be in full possession of his wits. He talked earnestly, had figured out precise specifications, and knew exactly what he wanted. The Goodyear people promised to make a square balloon of rubberized cloth. They delivered it a couple of months later to the Squash Court. It came neatly folded, but, once unfolded, it was a huge thing that reached from floor to ceiling.

The Squash Court ceiling could not be pushed up as the physicists would have liked. They had calculated that their final pile ought to chain-react somewhat before it reached the ceiling. But not much margin was left, and calculations are never to be trusted entirely. Some impurities might go unnoticed, some unforeseen factor might upset theory. The critical size of the pile might not be reached at the ceiling. Since the physicists were compelled to stay within that very concrete limit, they thought of improving the performance of the pile by means other than size.

The experiment at Columbia with a canned pile had indicated that such an aim might be attained by removing the air from the pores of the graphite. To can as large a pile as they were to build now would be impracticable, but they could assemble it inside a square balloon and pump the air from it if necessary.

The Squash Court was not large. When the scientists opened the balloon and tried to haul it into place, they could not see its top from the floor. There was a movable elevator in the room, some sort of scaffolding on wheels that could raise a platform. Fermi climbed onto it, let himself be hoisted to a height that gave him a good view of the entire balloon, and from there he gave orders:

"All hands stand by!"

"Now haul the rope and heave her!"

"More to the right!"

"Brace the tackles to the left!"

To the people below he seemed an admiral on his bridge, and "Admiral" they called him for a while.

When the balloon was secured on five sides, with the flap that formed the sixth left down, the group began to assemble the pile inside it. Not all the material had arrived, but they trusted that it would come in time.

From the numerous experiments they had performed so far, they had an idea of what the pile should be, but they had not worked out the details, there were no drawings nor blueprints and no time to spare to make them. They planned their pile even as they built it. They were to give it the shape of a sphere of about 26 feet in diameter, supported by a square frame, hence the square balloon.

The pile supports consisted of blocks of wood. As a block was put in place inside the balloon, the size and shape of the next were figured. Between the Squash Court and the near-by carpenter's shop there was a steady flow of boys, who fetched finished blocks and brought specifications for more on bits of paper.

When the physicists started handling graphite bricks, everything became black. The walls of the Squash Court were black to start with. Now a huge black wall of graphite was going up fast. Graphite powder covered the floor and made it black and as slippery as a dance floor. Black figures skidded on it, figures in overalls and goggles under a layer of graphite dust. There was one woman among them, Leona Woods; she could not be distinguished from the men, and she got her share of cussing from the bosses.

The carpenters and the machinists who executed orders with no knowledge of their purpose and the high-school boys who helped lay bricks for the pile must have wondered at the black scene. Had they been aware that the ultimate result would be an atomic bomb, they might have renamed the court Pluto's Workshop or Hell's Kitchen.

To solve difficulties as one meets them is much faster than to try to foresee them all in detail. As the pile grew, measurements were taken and further construction adapted to results.

The pile never reached the ceiling. It was planned as a sphere 26 feet in diameter, but the last layers were never put into place. The sphere remained flattened at the top. To make a vacuum proved unnecessary, and the balloon was never sealed. The critical size of the pile was attained sooner than was anticipated.

Only six weeks had passed from the laying of the first graphite brick, and it was the morning of December 2.

Herbert Anderson was sleepy and grouchy. He had been up until two in the morning to give the pile its finishing touches. Had he pulled a control rod during the night, he could have operated the pile and have been the first man to achieve a chain reaction, at least in a material, mechanical sense. He had a moral duty not to pull that rod, despite the strong temptation. It would not be fair to Fermi. Fermi was the leader. He had directed research and worked out theories. His were the basic ideas. His were the privilege and the responsibility of conducting the final experiment and controlling the chain reaction.

"So the show was all Enrico's, and he had gone to bed early the night before," Herbert told me years later, and a bit of regret still lingered in his voice.

Walter Zinn also could have produced a chain reaction during the night. He, too, had been up and at work. But he did not care whether he operated the pile or not; he did not care in the least. It was not his job.

His task had been to smooth out difficulties during the pile construction. He had been some sort of general contractor: he had placed orders for material and made sure that they were delivered in time; he had supervised the machine shops where graphite was milled; he had spurred others to work faster, longer, more efficiently. He had become angry, had shouted, and had reached his goal. In six weeks the pile was assembled, and now he viewed it with relaxed nerves and with that vague feeling of emptiness, of slight disorientation, which never fails to follow completion of a purposeful task.

There is no record of what were the feelings of the three young men who crouched on top of the pile, under the ceiling of the square balloon. They were called the "suicide squad." It was a joke, but perhaps they were asking themselves whether the joke held some truth. They were like firemen alerted to the possibility of a fire, ready to extinguish it. If something unexpected were to happen, if the pile should get out of control, they would "extinguish" it by flooding it with a cadmium solution. Cadmium absorbs neutrons and prevents a chain reaction.

Leona Woods, the one girl in that group of men, was calm and composed; only the intensity of her deep dark eyes revealed the extent of her alertness.

Among the persons who gathered in the Squash Court on that morning, one was not connected with the Met. Lab.—Mr. Crawford H. Greenewalt of E. I. duPont de Nemours, who later became the president of the company. Arthur Compton had led him there out of a near-by room where, on that day, he and other men from his company happened to be holding talks with top Army officers.

Mr. Greenewalt and the duPont people were in a difficult position, and they did not know how to reach a decision. The Army had taken over the Uranium Project on the previous August and renamed it Manhattan District. In September General Leslie R. Groves was placed in charge of it. General Groves must have been of a trusting nature: before a chain reaction was achieved, he was already urging the duPont de Nemours Company to build and operate piles on a production scale.

In a pile, Mr. Greenewalt was told, a new element, plutonium, is created during uranium fission. Plutonium would probably be suited for making atomic bombs. So Greenewalt and his group had been taken to Berkeley to see the work done on plutonium, and then flown to Chicago for more negotiations with the Army.

Mr. Greenewalt was hesitant. Of course his company would like to help win the war! But piles and plutonium!

With the Army's insistent voice in his ears, Compton, who had attended the conference, decided to break the rules and take Mr. Greenewalt to witness the first operation of a pile.

They all climbed onto the balcony at the north end of the Squash Court; all, except the three boys perched on top of the pile and except a young physicist, George Weil, who stood alone on the floor by a cadmium rod that he was to pull out of the pile when so instructed.

And so the show began.

There was utter silence in the audience, and only Fermi spoke. His gray eyes betrayed his intense thinking, and his hands moved along with his thoughts.

"The pile is not performing now because inside it there are rods of cadmium which absorb neutrons. One single rod is sufficient to prevent a chain reaction. So our first step will be to pull out of the pile all control rods, but the one that George Weil will man." As he spoke others acted. Each chore had been assigned in advance and rehearsed. So Fermi went on speaking, and his hands pointed out the things he mentioned.

"This rod, that we have pulled out with the others, is automatically controlled. Should the intensity of the reaction become greater than a pre-set limit, this rod would go back inside the pile by itself.

"This pen will trace a line indicating the intensity of the radiation. When the pile chain-reacts, the pen will trace a line that will go up and up and that will not tend to level off. In other words, it will be an exponential line.

"Presently we shall begin our experiment. George will pull out his rod a little at a time. We shall take measurements and verify that the pile will keep on acting as we have calculated.

"Weil will first set the rod at thirteen feet. This means that thirteen feet of the rod will still be inside the pile. The counters will click faster and the pen will move up to this point, and then its trace will level off. Go ahead, George!"

Eyes turned to the graph pen. Breathing was suspended. Fermi grinned with confidence. The counters stepped up their clicking; the pen went up and then stopped where Fermi had said it would. Greenewalt gasped audibly. Fermi continued to grin.

He gave more orders. Each time Weil pulled the rod out some more, the counters increased the rate of their clicking, the pen raised to the point that Fermi predicted, then it leveled off.

The morning went by. Fermi was conscious that a new experiment of this kind, carried out in the heart of a big city, might become a potential hazard unless all precautions were taken to make sure that at all times the operation of the pile conformed closely with the results of the calculations. In his mind he was sure that if George Weil's rod had been pulled out all at once, the pile would have started reacting at a leisurely rate and could have been stopped at will by reinserting one of the rods. He chose, however, to take his time and be certain that no unforeseen phenomenon would disturb the experiment.

It is impossible to say how great a danger this unforeseen element constituted or what consequences it might have brought about. According to the theory, an explosion was out of the question. The release of lethal amounts of radiation through an uncontrolled reaction was improbable. Yet the men in the Squash Court were working with the unknown. They could not claim to know the answers to all the questions that were in their minds. Caution was welcome. Caution was essential. It would have been reckless to dispense with caution.

So it was lunch time, and, although nobody else had given signs of being hungry, Fermi, who is a man of habits, pronounced the now historical sentence:

"Let's go to lunch."

After lunch they all resumed their places, and now Mr. Greenewalt was decidedly excited, almost impatient.

But again the experiment proceeded by small steps, until it was 3:20.

Once more Fermi said to Weil:

"Pull it out another foot"; but this time he added, turning to the anxious group in the balcony: "This will do it. Now the pile will chain-react."

The counters stepped up; the pen started its upward rise. It showed no tendency to level off. A chain reaction was taking place in the pile.

Leona Woods walked up to Fermi and in a voice in which there was no fear she whispered: "When do we become scared?"

Under the ceiling of the balloon the suicide squad was alert, ready with their liquid cadmium: this was the moment. But nothing much happened. The group watched the recording instruments for 28 minutes. The pile behaved as it should, as they all had hoped it would, as they had feared it would not.

The rest of the story is well known. Eugene Wigner, the Hungarian-born physicist who in 1939 with Szilard and Einstein had alerted President Roosevelt to the importance of uranium fission, presented Fermi with a bottle of Chianti. According to an improbable legend, Wigner had concealed the bottle behind his back during the entire experiment.

All those present drank. From paper cups, in silence, with no toast. Then all signed the straw cover on the bottle of Chianti. It is the only record of the persons in the Squash Court on that day.

The group broke up. Some stayed to round up their measurements and put in order the data gathered from their instruments. Others went to duties elsewhere. Mr. Greenewalt hastened to the room where his colleagues were still in conference with the military. He announced, all in one breath, that Yes, it would be quite all right for their company to go along with the Army's request and start to build piles. Piles were wonderful objects that performed with the precision of a Swiss watch, and, provided that the advice of such competent scientists as Fermi and his group were available, the duPont company was certainly taking no undue risk.

Arthur Compton placed a long-distance call to Mr. Conant of the Office of Scientific Research and Development at Harvard.

"The Italian Navigator has reached the New World," said Compton as soon as he got Conant on the line.

"And how did he find the natives?"

"Very friendly."

Here the official story ends, but there is a sequel to it, which started on that same afternoon when a young physicist, Al Wattemberg, picked up the empty Chianti bottle from which all had drunk. With the signatures on its cover, it would make a nice souvenir. In subsequent years Al Wattemberg did his share of traveling, like any other physicist, and the bottle followed him. When big celebrations for the pile's tenth anniversary were planned at the University of Chicago, the bottle and Al Wattemberg were both in Cambridge, Massachusetts. Both, Al promised, would be in Chicago on December 2.

It so happened, however, that a little Wattemberg decided to come into this world at about that time, and Al could not attend the celebrations. So he shipped his bottle, and, because he wanted to make doubly sure that it would not be broken, he insured it for a

August 2nd, 1939

F.D. Roosevelt,
President of the United States,
White House
Washington, D.C.

Sir:

Some recent work by E.Fermi and L. Szilard, which has been communicated to me in manuscript, leads me to expect that the element uranium may be turned into a new and important source of energy in the immediate future. Certain aspects of the situation which has arisen seem to call for watchfulness and, if necessary, quick action on the part of the Administration. I believe therefore that it is my duty to bring to your attention the following facts and recommendations:

In the course of the last four months it has been made probable - through the work of Joliot in France as well as Fermi and Szilard in America - that it may become possible to set up a nuclear chain reaction in a large mass of uranium,by which vast amounts of power and large quantities of new radium-like elements would be generated. Now it appears almost certain that this could be achieved in the immediate future.

This new phenomenon would also lead to the construction of bombs, and it is conceivable - though much less certain - that extremely powerful bombs of a new type may thus be constructed. A single bomb of this type, carried by boat and exploded in a port, might very well destroy the whole port together with some of the surrounding territory. However,

I understand that Germany has actually stopped the sale of uranium from the Czechoslovakian mines which she has taken over. That she should have taken such early action might perhaps be understood on the ground that the son of the German Under-Secretary of State, von Weizsäcker, is attached to the Kaiser-Wilhelm-Institut in Berlin where some of the American work on uranium is now being repeated.

Yours very truly,

A. Einstein

(Albert Einstein)

NO SMOKING IN THIS ROOM

The First Atomic Pile under Construction in the Squash Court: Chunks of Uranium Are Imbedded in the Graphite Bricks

thousand dollars. It is not often that an empty bottle is considered worth so much money, and newspaper men on the lookout for sensation gave the story a prominent position in the press.

A couple of months later the Fermis and a few other physicists received a present: a case of Chianti wine. An importer had wished to acknowledge his gratitude for the free advertisement that Chianti had received.

(20)

SITE Y

A girl who worked in the transportation office at the Met. Lab. sent me the railroad tickets by a messenger boy. She had just called me on the telephone:

"I have got you a drawing-room on the 'Chief,' " she had said with well-deserved pride—reservations for civilians were hard to obtain in that summer of 1944. "You know you are to get off at Lamy, don't you? Somebody will meet you there."

"And if he does not?" I had wanted to ask, but she had hung up.

Giulio saw the slips of paper in my hands.

"Are they the tickets to Site Y? I want to see them. Let me have them!" he said, jumping excitedly up and down.

"Keep quiet. There is nothing to see on the tickets," I answered. My own excitement had vanished at the prospect of the vague connection with an unknown person at Lamy, New Mexico. Should we miss *him,* or *her* for all I knew, we would reach Santa Fe somehow. But, once there, could we go around asking people to help us find the most secret place in the United States?

Perhaps I would have done better to wait for Enrico. We had adjusted our plans to his. The best time for him to move from Chicago to Site Y would be the middle of August, he had said, and I had sublet our home, given notice to the maid, engaged movers. Then, early in August, Enrico had been called elsewhere. Top-secret mission. I could write to him at Richland, Washington. Of course, I did not know that near the open town of Richland was Hanford, where on a closed area the duPont company was building atomic piles, nor that Enrico was to assist the duPont people in starting operations.

"You can wait for me in Chicago, or you can go ahead to Site Y with the children," Enrico had said.

"Why can't we go now?" Nella had asked.

"I can't wait any more," Giulio had added.

"When will you come back?" I had asked Enrico.

"I haven't the faintest idea," he had answered.

Both children had looked at me with long, worried faces. Perhaps because of its secrecy, Site Y had caught their imagination. It had caught mine, too, I must admit.

Site Y was the place in the middle of some strange wilderness, where many of our friends had disappeared; where several of the European-born were unhappy because living inside a fenced area reminded them of concentration camps; the place whose real name was unknown to me, but where I was going to have an apartment with windows that a physicist had measured with great precision, so that I might purchase curtains in Chicago. Site Y was the place where a pair of hiking shoes would be a godsend, rationing permitting; and another godsend would be an electric roaster, because there was no gas; where children were allowed to roam around at their will, as they could not go beyond the fence anyhow; where censorship of private mail had been announced to residents a short time *after* a few enterprising physicists had carried out experiments which proved that letters were opened. Emilio Segré, who had lived at Site Y with his family since the summer of 1943, once went on a trip into the outside world. He wrote a letter to his wife and put a strand of hair in it, but when she opened the letter, the hair was gone.

At Site Y, I would certainly be happy, according to a young man who chewed pensively at a pipe and whom I had never met before. He came to see us in Chicago to explain what he could about Site Y, where he was the director of the laboratories. His name was Robert Oppenheimer.

These were the facts that so appealed to the children and to me.

"Can't we go now?" the children had pleaded.

That August was hot and unpleasant in Chicago. I had made up my mind.

"We'll go without you," I had said resolutely to Enrico.

Now that we were ready to leave and Enrico was away, the fear

of being stranded in Santa Fe increased, and I felt a growing un-
easiness. It was unexpectedly relieved the night before our departure
when Arthur Compton, who had been at Site Y before, called to say
he was going there for a visit and would ride our same train. And
could Giulio use a pair of fishing boots that his son John had outgrown but not worn out? There were many streams in the mountains
near Site Y and good trout, he said, and Betty—Arthur's efficient
wife—thought we might not have a spare coupon for boots.

That Betty was decidedly an efficient wife, I realized the next day,
when Arthur came to our drawing-room on the train with the boots
in his brief case and "Betty's order to ask you all for dinner . . .
and her permission," as he swiftly added.

On that same train I had the pleasant surprise of seeing another
of our friends, Harold Urey, whom I had left in Leonia over two
years before. It would be more accurate to say that through the
open door of a roomette I saw a tired-looking man who looked like
Harold Urey, stretched on the divan, absorbed in who knows what
thoughts and what deep concern. Nella and Giulio, whom I dispatched in reconnaissance, reported that yes, that man must be
Mr. Urey. Harold was overworked and tired and looking older
than his age all during the war years; he recovered only when he
could put his mind at rest about the war and his wartime duties.

I could not go up to him and say: "Aren't you Harold Urey?"
Not even: "Hello, Harold!"

Most of the important scientists traveled under false names in
those days. His might have changed into Hugh Ulman or Hiram
Upton, for the Army, who was responsible for the changes, had
imagination and saved only the initials. Enrico was Eugene Farmer
when he traveled, and Arthur Compton had two names, Mr. Comas
and Mr. Comstock, one for the East and one for the West. Once
he was napping on a flight to California from New York when the
hostess woke him up to ask his name.

"Where are we?" Compton inquired by way of reply, and looked
out of the airplane window.

So I did not dare speak to Harold Urey. But after a while he
emerged from his thoughts and saw us. We talked at length about
our families, the friends we have in common, about the latest news
of the war, the Allied victories in France—our troops were rapidly

approaching Paris—but neither of us mentioned our destination or the purpose of our trip.

We all got off the train at Lamy.

I had hardly set foot on the platform when a blond soldier walked up to me.

"Are you Mrs. Farmer?" he asked.

"Yes, I am Mrs. Fermi."

"I was told to call you Mrs. Farmer," he said mildly, but there was reproof in his blue eyes. Among the instructions I had received in Chicago, none indicated that I was to use Enrico's new name.

The soldier motioned us to a GI car and drove us the sixty-odd miles to our destination. Compton and Urey were whisked away in another car. They were to attend a meeting at Site Y.

The story of Site Y starts in the fall of 1942, almost two years before I moved there with my children.

As far back as December, 1941, the top policy group of the Uranium Project had expressed the opinion that the Army should take over the entire project, once the phase of actual construction of bombs was reached.

On August 13, 1942, a special district was formed in the Army Corps of Engineers to carry out atomic work. Its name, "Manhattan District," was intended to conceal all connection with atomic research. On the following September 17, Secretary of War Henry L. Stimson placed Brigadier General Leslie R. Groves in charge of the Manhattan District. It was decided at the same time that work should be greatly expanded as soon and as fast as possible.

General Groves set himself to his task. He planned work on the atomic bomb even before the pile experiment on December 2, 1942, had definitely proved that release of atomic energy was feasible. He chose sites for new laboratories and production plants.

The duPont company was going to erect big piles for the production of plutonium at Hanford, Washington, on the Columbia River. Separation of very fissionable from less fissionable uranium was to be carried out at Site X, a tract of land at Oak Ridge, Tennessee. Study of all problems related to design and construction of an atomic bomb, the most secret part of the project, had to be furthered in a spot even more secluded than the other two.

In the search for such a place General Groves was helped by Professor Robert Oppenheimer, or "Oppie," as his friends called him. Oppie's family owned a ranch in the Pecos Valley, on the eastern side of the Sangre de Cristo Mountains. Oppie was thoroughly acquainted with New Mexico. He suggested to General Groves a small boys' school on a lonely mesa by the Los Alamos Canyon, on the slopes of the Jémez hills.

In Oppie's opinion the school and its grounds filled General Groves's requirements for the new laboratories. A narrow mountain road, passable, if not good, connected the school to the main highway from Taos to Santa Fe. The existence of this road would help surmount the difficulties of transportation. By way of Española, Santa Fe was forty-five miles away, and the nearest railroad was at a distance of over sixty miles. The mesa by the Los Alamos Canyon was undoubtedly a secluded spot. The school buildings could accommodate the first scientists during the phase of construction. There was no danger of running out of space for expansion, for proving grounds, for erecting large machinery: as far as the eye could see, there were only pines and sand.

General Groves was much interested in the school, and with Oppie he took the train to Santa Fe.

The valley of the Rio Grande River north of Santa Fe is a wide basin that was a lake in some remote past. It is closed by the softly jagged range of the Sangre de Cristo Mountains to the east and the green domes of the Jémez hills to the west and northwest.

The bottom of the basin is hot and barren: sand, cacti, a few piñon trees hardly rising above the ground, and space, immense, transparent, with no fog or moisture. Only along the curves of the Rio Grande and on thin strips bordering the creeks that come from the mountains is the land green, fertile, and can it give life. There the ancient Indian pueblos and Spanish villages rest under the shade of long-lived trees, seek the gifts that the river bestows upon the land, and accept the sterility of the desert, the inexorability of the New Mexican sun.

The Jémez hills rise above a high plateau, cut by deep canyons in a series of parallel mesas like the teeth of a giant comb. The canyons, the arroyos, the gullies, were formed by the erosive action of water, but they are mostly dry. They fill only when a violent storm in the mountains and hills suddenly hurls down a wall of water. It rushes forward, dangerous, impetuous, furious. It drags along the clay that it steals from the sides of the mesas and pours it into the Rio Grande. There the violence of the water abates; the river flows in slow waves, thick and muddy; and if the storm comes from those mesas to the north, where the soil is red, the river also becomes gloriously red, like blood in an open artery.

The Los Alamos Ranch School for Boys was built on a mesa high above the valley, with steep, straight sides streaked with gold and red, with a pale-green top, the color of pine trees covered by the dust that the wind whirls up from the desert below.

General Groves and Oppie went to see the school. The school principal must have been surprised by the strangely assorted pair: a slender intellectual with rounded shoulders and narrowed eyes, who acted as a guide to a burly Army officer, straight, direct of manner, with authority in his voice. The principal must have been even more surprised at his visitors' requests. The school must be closed. The Army wished to buy it for secret work.

The Manhattan District purchased the school in November, 1942. Oppie was to direct the future laboratories, and General Groves asked him how many houses would be needed. Oppie expected to gather some thirty scientists and their families, perhaps a hundred persons in all.

Oppie turned out to be a marvelous director, the real soul of the project. In his quiet, unobtrusive way, he kept informed about everything and in touch with everyone. His profound understanding of all phases of research—experimental, theoretical, technical—permitted him to co-ordinate them into a coherent whole and to accelerate the work. He carried the burden of his responsibilities with an enthusiasm and a zeal bordering on religiosity. But in predicting the size of Site Y he did not do so well.

The Special Engineer Detachment started construction in January, 1943. They built homes. They built laboratories following vague directives of scientists who could not explain the kind of work they would pursue in them. As the Engineers completed the first order of buildings, the scientists asked for more; and then for even more. Thus a city grew at 7,200 feet above sea-level.

Into that city went scientists from all parts of the United States and England, to disappear from the world. For two and a half years the city was not marked on the maps, had no official status, was not part of New Mexico, its residents could not vote. It did not exist. That city was Los Alamos to those living there, Site Y to the few outsiders who knew about its existence, Post Office Box 1663 to correspondents and friends of the inhabitants.

The influx of new families on the mesa never ceased, and building went on at a feverish pace, invariably lagging behind the increase in population. At the end of the war, when the first statistics were published, the Los Alamos population was 6,000. Ten years after its birth Los Alamos had 12,000 residents and a housing shortage.

Thus building activities never ceased. When we arrived in Los Alamos, in August, 1944, we found the confusion and the disorder that always accompany a fast pace of construction.

Around the few original houses of the school, barrack-like buildings seemed to have been scattered at random. They stood at strange angles on streets without names that loafed aimlessly about the mesa and drew intricate patterns over it. The buildings were all alike, all painted green, inconspicuous among green pines, against the green background of the hills. One easily got lost in the uniformity of the houses and the intricacy of the streets with no names, but one could regain his orientation by the single landmark, the tall water tower on the highest part of town.

Long afterward I recognized a logical design in the layout of houses: they were set diagonally to the streets, to utilize the ground best, leaving, however, sufficient space between them to keep fire hazards low. All houses were made of wood.

Buildings under construction emerged from thick seas of mud. There is never much protective vegetation on a high mesa, and even the little there is dies out when construction is under way. During the rainy season in the summer, the downpours turned the clay of the soil into slippery glue that stuck to the shoes and then hardened into heavy soles. When winter came, the snow, melting under the balmy rays of the midday sun, again turned the soil into mud. Construction materials and felled trees were piled along the sides of rutted roads where bulldozers, cranes, and trucks sped blindly away as if they were the masters of the place.

Along the Los Alamos Canyon a strip of mesa was fenced off with chicken wire. Behind it was the Technical Area, where only persons with special badges could gain admittance. The main town entrance, the East Gate, led into the desert to Española and Santa Fe. Through the West Gate, open to civilians during certain hours only, one reached the mountain country, the fishing streams, the ski slopes, and the woods—woods of blue spruce, of Ponderosa pines, and of the aspens that turn yellow in the fall and cover the hills with foils of gold. Through the West Gate one reached the spectacular Valle Grande, an immense basin lined with a single-piece carpet of green grass; it had been a volcano crater in times past, and the Jémez hills were its borders.

No matter by what gate you left town or came into it, you had to show your pass to the MP's on guard. But all the children in Los Alamos knew the holes in the barbed-wire fence that inclosed the town, and they could act as guides to the adults. A second fence went round the first a few yards farther, and whether there were holes in it I have never known, for my own guide, Giulio, never took me there.

We were assigned apartment D in building T-186, one of a dozen identical four-apartment houses down a street that started near the water tower on the summit of town, sloped leisurely toward the virgin country, and faded away into it.

The apartment was small but adequate and comfortable. In its three bedrooms were army cots on which their previous occupants, boys in the armed forces, had carved their names and ranks. Sheets and blankets were stamped USED in big black letters that shocked us greatly until we realized that they stood for United States Engineer Detachment. Everything provided by the project was either USED or GI, even light bulbs and floor mops. But through the three contiguous windows of our living-room I could see the round green tops of the Jémez hills slanting down against the sky, as in a three-panel picture by an old master. There were no man-made marks on the hills, and I could call them mine.

At night no street lights spoiled the darkness, and if we saw the black shadows of the pines on the white snow in winter, we knew the moon was out.

We had hardly taken possession of our apartment and were busily unpacking our suitcases when there was a knock at the kitchen door. Kitchens were on the street side of the buildings, and no one cared to cross muddy yards to reach the living-room entrances. There were no telephones in our homes, and knocks at the door were frequent.

A man and a woman walked in, the woman taller, with stronger features and bigger bones, with apparently greater assurance, than the small man of modest countenance in whose pale face one noticed only the intelligent eyes. When the woman spoke in answer to my puzzled look, her voice struck me by its resonance and foreign accent, more marked than any I had heard before.

"Don't you remember us? We are Peierls. We met in Rome in 1933. Now we live in apartment below, apartment B in same building." She skipped all articles from her otherwise fluent sentences.

Now I remembered. Rudolf Peierls had been a foreign fellow in the physics department in Rome for several months, soon after he had emigrated to England from Germany, where he was born. Then I had known his wife Genia well.

Rudolf explained he had come from England as a member of the British Mission, a group of British scientists who were to help their American colleagues at Los Alamos. They had arrived a few days before us, and their two children had joined them from Canada where they had been evacuated when the bombing of London had started.

We hailed the act of Fate that placed our two families one on top of the other. Our children became friends. Genia had not gone through the process of absorbing American ways from her children, she had not been exposed to their interpretation of democracy and freedom or to their criticism of European parents. She was still raising them in the strict, British manner. Her loud scoldings carried through the flimsy floors of our house and delighted us by their quaintness. We Europeans noticed oddities of speech in others but were insensitive to our own.

On one of our first afternoons in Los Alamos, Genia Peierls came to propose a picnic in Frijoles Canyon.

"You must take car," she asserted. Members of the British Mission were far more austere than the Americans: they cooked in the scanty supply of GI pots and pans, they owned no cars. "We'll be large group. Mind me, you'll always be in large groups here. It's merrier. Today all cars will be filled up. Persons who come don't all have spare coupons. You can drive car. It's only eighteen miles through Western Gate, and we'll go after five, so Western Gate will be open and we'll avoid long detour through Eastern."

I was hesitant. Frijoles Canyon contains ruins of the oldest Indian pueblos in that region and some well-preserved cave dwellings. I had never seen either. On the other hand, I am a timid driver, I had never driven in rough country, I mistrusted the road and the several-hundred-foot drop I knew it took from the edge of a mesa down to the bottom of the canyon.

"Mind me, Laura," Genia said. "Somebody will drive your car. All have driving licenses in group." It was impossible to resist Genia's spirits and her enthusiasm for any sort of action.

I found at the steering wheel of my car an attractive young man, slim, with a small, round face and dark hair, with a quiet look through round eyeglasses. He could not have been much over thirty years of age. I tried to make friends with him and asked him some questions, which he answered sparingly, as if jealous of his words. Perhaps he was absorbed by the driving. He was not a good driver and wriggled the car jerkily on the narrow road. He must have been nervous.

I extracted little information from him: that he was born in Germany; that he had been a refugee in England and had British citizenship; that he was a member of the British Mission and had recently arrived in Los Alamos. My attempts at friendliness seemed lost on him, although he was extremely polite and of refined and cultured manners. I never understand names when I first hear them and had not caught his. When we said goodbye in front of my home after the picnic, I asked him to repeat it. He was Klaus Fuchs.

Even as he spoke to me, he was leading a double life—that of highly competent and appreciated physicist among friendly colleagues and that of spy. He was giving secret information to the Russians on the progress of the atomic bomb. He had aroused no suspicion. When in 1950 Fuchs confessed his spying activities, he claimed he had some sort of split personality and could keep his friends and his political ideals in separate compartments. He said: "It had been possible for me in one half of my mind to be friends with people, to be close friends, and at the same time to deceive and to endanger them."

In Los Alamos we all trusted him and saw him frequently, for he attended many of the numerous parties that went on constantly. There was little else to do at night: men could not talk of their work with their wives; the only places of entertainment were the movies. So we had frequent parties, and Fuchs came often. He seemed to enjoy himself, played "murder" or charades with the others, and said only a few words. We all thought him pleasant and knew nothing about him.

Early in 1950 he made a full confession under very little pressure, almost of his own free will. There was hardly any evidence against him, but he was having increasing scruples about Russia's true aims and the sincerity of communism. Enrico and I followed the investigation and the trial through the papers. One fact struck us as peculiar and hard to believe: that Fuchs was never conscious of the full import of his behavior. He felt guilty of deceit toward his friends, not of betrayal toward the country he had elected for his own and had sworn allegiance to. He did not expect to be held responsible to mankind for the danger he had brought about. He had not even foreseen the legal consequences of his confession, the judge's sentence, his prison term. He was aware, he said to the British investigator, that it might be better if he resigned from his post at Harwell, once his past was disclosed.

Harwell is the British counterpart of Los Alamos, the place where secret atomic research is carried on, where Fuchs held a prominent, directive position. Although he had not applied the word "spy" to himself, still he realized that a man with a record like his could hardly be trusted with important secret work. But he was certain he would have no difficulty obtaining a teaching position at one of the English universities!

Fuchs's intelligence was far above the average; his judgment below, probably distorted by the circumstances of his life. Both in Germany, where he had been in the underground movement, and later in England he had heard much praise of disloyalty against the established government of Hitler. It was easy to jump to the conclusion that the individual's judgment is basically sounder than that of the state, of established society.

As a governing body, even England committed actions that wounded individuals, that wounded Klaus Fuchs. In May, 1940, all enemy aliens were interned regardless of their political belief. Although a refugee from Nazi persecution, Fuchs was sent to a camp on the Isle of Man, then to Canada. England could not take risks with aliens, for she was facing an emergency that was perhaps the gravest since the times of the Invincible Armada. The phony war was over, and Hitler had swept through Denmark, Norway, Holland, Belgium, and Luxembourg. France was invaded. The struggle for survival was on in England.

In January, 1941, Fuchs was released from the internment camp in Canada, and he returned to England. Great Britain had signed a twenty-year mutual-aid pact with Russia on the previous July. On the following June Hitler launched his offensive against Russia, and in the Communist defenders came to rest all hopes of a Nazi defeat. At this historical moment Fuchs did what his judgment told him to do. He offered his help to the Russians. To go against laws set by a government, even the British that had granted him political asylum when he needed it, was not morally wrong in his mind.

(21)

A BODYGUARD AND A FEW FRIENDS

I had been in Los Alamos some three weeks when Enrico arrived from Hanford with his bodyguard, John Baudino. Yes, Enrico did have a bodyguard!

General Groves keenly felt his responsibility toward the men of science who had suddenly assembled under his wing. He designated some half-dozen of his most valuable and vulnerable charges to be protected by a member of the Army Intelligence in plain clothes.

Perhaps because Enrico was from an enemy country from which sabotage and retaliations might be expected, General Groves included him among the scientists needing special protection. Moreover, he discouraged him from flying, which he considered more hazardous than a quiet train ride.

John Baudino had made his appearance in our lives early in 1943, while we were still in Chicago. One evening I went to answer our doorbell and found myself face to face with a big man who entirely filled the doorway. The light on the front porch shining from behind him made him look like a huge, looming apparition. In a deep voice with no harsh tones, the apparition shyly asked me to tell Dr. Fermi he would wait for him outside.

The rules General Groves had set could have been chosen by a wise mother for her teen-age daughter. Enrico was not to walk by himself in the evening, nor was he to drive without escort to the newly built Argonne Laboratory twenty miles from our home. The pile assembled under the West Stands had been moved to Argonne, as that laboratory was usually called, and in the spring of 1943 Enrico drove there almost daily.

Enrico is an individualist and values his independence above all.

He will never follow a guide on a mountain climb, and whatever height or peak he can attain must be his own attainment. He prefers walking miles rather than being driven. He was never in a car pool, not even when gas was rationed, because conforming to other people's habits went against his nature.

I was well aware of Enrico's idiosyncrasies and felt strong misgivings about his relationship with a bodyguard. I was wrong. Enrico did not resent Baudino's company at all but grew increasingly fond of him.

Baudino was an Illinois-born son of immigrants from northern Italy, and in peacetime he practiced law. Whether his Italian origin had influenced his designation as Enrico's bodyguard I was never able to establish. At the end of the war, when true and false stories about atomic scientists were busily divulged by the press, a magazine asserted that Baudino was chosen as bodyguard and interpreter because Enrico had a scanty knowledge of the English language. The truth is that Baudino knew only a few words of Italian and he and Enrico always spoke in English.

Baudino's size and his powerful build undoubtedly qualified him as a guard. He looked as if he had sufficient strength to wring the neck of any evil-minded spy or saboteur. In reality, he was kind and always ready to help others; Enrico appreciated his cheerful disposition, his eagerness to please, his endless store of entertaining tales.

It should have been his task to drive Enrico to Argonne, but Enrico's fondness for Baudino, deep and sincere as it was, proved insufficient to overcome his reluctance to let others drive him. So Enrico sat at the steering wheel, and Baudino rode by him, a hand on his gun and his face turned back toward the cars behind. Baudino kept a mental record of all license plates so he would spot at once any car that might be following with suspicious insistence.

At Argonne, while waiting for Enrico, he gave legal advice to anyone in need of it, helped fill income tax returns, and was ready for all sorts of odd jobs. In return, he asked explanations of physics. Soon he could operate the pile and had learned so much physics that Enrico said he deserved an honorary degree. He never got it.

Baudino chaperoned Enrico during his frequent trips out of town, and then he was invaluable. Before priority for travel was accorded to scientists, wartime conditions made it almost impossible to obtain reservations at the last moment. If at the ticket office of a railroad station he was told that not a single Pullman accommodation was available, Baudino would walk up to the conductor of the first train to Chicago. He would square his shoulders, raise himself to his full portly stature, bang his fist on whatever banging surface was at hand, and state boldly:

"It is imperative that we leave immediately for Chicago."

Train conductors are frail old men more often than not. Faced with Baudino's bulk and reliance, they could do little but produce the required accommodations.

In the winter of 1944 Enrico went to Site Y on a business trip. One Sunday morning Emilio Segré and Hans Bethe, a German-born physicist, suggested a skiing trip. A question arose. Should Emilio or Hans waste precious gas allowance and drive their cars, or could they take the military car at Enrico's disposal during his stay? This car was to be used for strictly business purposes, and a skiing excursion could be hardly called business, as Enrico sadly conceded.

"It is not business for you, and therefore you should not take the car," clarified lawyer Baudino, who was anxious not to miss the trip. "But should you decide to go, it would be my business to accompany you. Hence I can take the car."

The story does not say what car they drove in the end. Anyhow they went. Baudino must have blamed himself for having encouraged the expedition: by the end of the day he was so completely exhausted that Enrico had to carry his gun. An Illinois boy out on his first mountain experience was no match for Bethe, Segré, and Fermi, who had tramped the snow of the Alps.

Not that the others could be called champions. Of Bethe's skiing I know nothing at first hand, but I assume that he would tackle it with the same slow steadiness with which he hiked in summer. Enrico comes down an incline crouching on his skis to keep his center of gravity as low as he can and to avoid falling, a posture hardly considered in good style. And about Emilio I heard certain stories of black-and-blue marks, which his friends were invited to see in the privacy of his room, that cast some doubt upon his skill.

Policy-making in Los Alamos: Lawrence, Fermi, and Rabi

Niels Bohr on Sawyer Hill

Los Alamos: A House in "Bathtub Row"

After the War the Old Green Houses Were Painted White
and the Streets Were Given Names

If Baudino was unhappy about this excursion, he did not grumble. He never grumbled. The discomforts of frequent journeys, which irked Enrico, were a pleasure to him. Enrico never traveled enough, according to Baudino, who urged him to take more trips. Life was dull otherwise, especially after Enrico settled in Los Alamos, a well-guarded place where he was permitted to roam around even at night without his bodyguard.

Despite rumors that spread unaccountably among our friends in the wide world outside our mesa, Baudino never slept in Enrico's room in Los Alamos. He had his own house, and there his wife and baby daughter joined him. Because Enrico did not travel with sufficient frequency to keep him busy, Baudino was given a job in the Security office, from which he was excused every time Enrico left the mesa and needed protection.

Segré and Bethe, the two men who had gone skiing with Enrico and Baudino, were on the mesa when I arrived. Although I had seen Emilio Segré a couple of times in the United States, I still thought of him as he had been in 1935, before he had left Rome for Palermo: a lean, dark-haired young man who cast incinerating glances round every time his susceptible feelings were hurt. But Segré had mellowed with age. His hair was graying on the temples, his waist had filled out considerably, and he was a respectable-looking married man of thirty-nine, the father of two children. He was friendly and obliging as never before.

"While the Pope is not here, I'll bring you your mail," he said to me. Enrico had never ceased to be "the Pope" to him.

"Why should *you* bring my mail? Isn't it delivered?"

"There is no home delivery. All mail goes to the Tech Area and there it stops."

Segré kept my automatic washer and others in working condition. There was no commercial service for electric appliances on the mesa, and it was hard to obtain a pass for a repairman from Santa Fe.

Emilio was not the only Italian friend whom I found in Los Alamos; neither was he the only one who pressed me to accept his services. In reality, Segré brought me my mail every other day, while the rest of the time I received it from the hands of another friend, Bruno Rossi. I had suggested this arrangement in an effort to prevent the two men's rivalry in being of assistance to me from resulting in a fight.

We had known Bruno Rossi in Italy for many years, but we had not become close friends because we had never lived in the same city. Bruno had taught in Florence first, then he had obtained the chair of experimental physics in Padua. There he was in the summer of 1938, when, after the *Anschluss*, the effects of Mussolini's enslavement to Hitler caused many Italians to turn their hopes on distant countries. Bruno left Italy with his recently acquired wife, spent a few months in England, and then landed in the United States, the only land of promise left to men of science.

We saw the Rossis when, having just arrived in this country, still bewildered, they looked at the new world around them with doubtful but hopeful eyes. Bruno, a few years younger than Enrico, was a quiet man, rather silent and shy. He let his lively wife do all the talking, and he was happy if he could withdraw into the background. After a couple of years in Chicago, Rossi took a position at Cornell University, and from there he went later to Los Alamos.

I knew that, despite his self-effacing manners, Rossi was a man of great intelligence who had gained an outstanding reputation in his field of cosmic rays. To study cosmic radiation, Bruno had climbed mountains, flown on dirigibles, sent balloons up into the air. All this made sense to me, for I was aware that cosmic rays come from interplanetary spaces and are stronger, the smaller the path they have traveled in the earth's atmosphere. In Los Alamos, however, Bruno worked at the very bottom of a deep canyon, by the straight side of our mesa. On the edge of the mesa, overlooking Rossi's canyon, there was a tiny hut known as "Segrè's castle."

What a cosmic-ray man like Rossi could do at such a low level and why Segrè's work should be located so much above Rossi's is one of the many Los Alamos mysteries that still haunt my mind.

At noon both Emilio and Bruno went back to the central part of the Tech Area and walked by the Tech Area Post Office. On alternate days they brought me my mail, until Enrico arrived and relieved them of this task.

Nobody was surprised at Rossi's kindness and **eagerness to** please, but this same attitude was startling in Emilio. He was so amiable, only the few persons who had known him well in Italy persisted in calling him the "Basilisk," rather to express affection than to comment on his temper. Hans Bethe was one of these.

Bethe had been the first foreign student of physics at the University of Rome. In this country we are used to continuity, and it has no meaning to say "the first foreign student." There have always been foreign students. But in Rome physics was not good enough to attract foreigners until Corbino gathered his group of "boys." As soon as his school became known outside Italy, foreign physicists began to visit Rome for reasons other than tourism.

Hans Bethe came from Germany to spend the spring of 1931 at the physics building in Rome. He had finished his formal training not long before, and when he came to see us in our home, he behaved like a boy in the presence of his teacher. He spoke little and slowly: I knew no German, and the conversation was laboriously carried on in English. His small eyes blinked often, and a heavy mass of chestnut hair stood straight up on his head. At dinner he ate slowly but steadily. Little by little, a mountain of spaghetti became a hill, disappeared altogether in his mouth between glassfuls of red wine.

Bethe spent more time thinking than talking. He was a theoretical physicist, and theoreticians are inclined toward cogitation. Bethe believed his thinking processes were too slow and was happy to learn speed from Fermi. He had been taught to tackle a problem all at once, as if it were a big dish of spaghetti, to put down all data into complicated formulas and to persevere. Perhaps in three months he would reach a solution. Enrico showed him how to reduce the problem to its essential elements, how to seek partial solutions at first, and how plain reasoning may replace rigorous mathematical deductions.

Bethe enjoyed Rome and came back in 1932. At that time he shared the great apprehension many Germans felt about the Nazi party and the increasing power of Hitler. The following year he left Germany, went first to England and then to the United States, where he took a position at Cornell University.

He went to Los Alamos in the spring of 1943, and there he was the leader of the theoretical physics division. Much of the success of the project is due to this division. There was no help from previous engineering experiences in constructing an atomic bomb. The choice of materials, the requirements of purity, the design, the critical size, and the properties of the bomb were all determined by theoretical studies based on results of experiments, sometimes on infinitesimal quantities of materials.

There were others in Los Alamos besides Segré, Bethe, and the Peierls with whom we could claim a friendship dating back from the times in Rome. One of them was Edward Teller.

Because I had never set eyes on a Hungarian before, I was much interested in Teller when I first met him. But I could not detect any intrinsic differences between him and men of other nationalities. His eyebrows were his outstanding characteristic. They were so thick, so bushy, they jutted out so much above his green eyes, that they looked like gables over the stained windows of some old church. When he was absorbed in thought, he thrust them up, and his face acquired a strange intensity. Surprise, bewilderment, shock, each gave a different shape to his eyebrows.

During his short stay in Rome, Teller had not talked physics with Fermi. The two had played ping-pong. In this game Teller was far more proficient and therefore a great challenge to Fermi.

In 1935 Edward Teller had been offered a position at George Washington University in Washington, D.C., and he had settled for good in the United States with his wife Mici. He and Enrico met again at Stanford University in California during the summer of 1937, while Enrico was lecturing there. Their friendship grew. Teller, his wife, and Enrico drove back east together. Years later the Tellers complained to me that they were obliged to submit to Enrico's whims, during this trip, and visit places whose bizarre names had caught his fancy. On Enrico's insistence they had taken long detours to see Canyon del Muerto and Meteor Crater in the desert of Arizona.

Shortly after we arrived in America to stay, Mici Teller yielded to a natural curiosity and came to see what Fermi's wife and children looked like. Her inspection of our family must have been satisfactory, and we became good friends. Between Enrico and Edward stronger ties developed than interest in ping-pong and in touring the Southwest. They found they were scientifically congenial. Enrico enjoyed Edward's original view of old problems and his numberless new ideas.

"That young man has imagination," Enrico used to say (Edward was his junior by several years, for he was born in 1908). "Should he take full advantage of his inventiveness, he will go a long way." His company was greatly stimulating.

Teller belonged to that small group of physicists whose efforts alerted President Roosevelt to the possible applications of atomic power in warfare.

The new field of physics revealed by the discovery of uranium fission was fertile ground for an imaginative mind. Edward gave himself to exploring it, to speculating far ahead of the few known facts. In 1939 and 1940 he often shared his thinking with Fermi, and they talked at length of the future atomic age. Fermi's mind works by comparisons and analogies. He could not visualize this hypothetical, nonexisting atomic bomb unless he found something to compare it with. But did anything comparable exist? The amount of energy released by each uranium atom undergoing fission is so great that a small quantity of uranium could produce a tremendous explosion. Were there sources of energy that much concentrated on our earth or in the universe?

"Perhaps meteorites," Edward suggested. When they had taken their trip together, in 1937, he and Enrico had visited Meteor Crater in Arizona, a large, deep depression in the desert due to the explosion of a meteorite. Would an atomic bomb have effects similar to those of a big meteorite? Meteor Crater is probably larger and deeper than what they thought a bomb could do. The size and depth of Meteor Crater may be due to the fact that the meteorite, endowed with tremendous speed, sank deeply into the ground before exploding. An atomic bomb would be likely to explode above ground. But in a general way the two phenomena might be comparable.

At about that time Teller and Fermi also discussed the possibility and the probable characteristics of a thermonuclear reaction: it is the principle on which the hydrogen bomb is based. It goes without saying that they never talked of these matters in my presence.

Teller's hands were not so deft as his thoughts. He happened to be in New York on Thanksgiving Day in 1940, and he asked himself to dinner at our house. We expected other company; our table would seat no more. Enrico went to the basement and started building an extension leaf. Edward rushed to his help. But soon he fell into eager talk, he agitated hammer and screw driver in the air to help expression, and he ended by placing a finger in the path of Enrico's saw. After I had dressed his wound, Enrico told him that the best help he, Edward, could give was to sit quietly and at a safe distance.

Despite his discussions with Fermi, Edward Teller had not joined the uranium project by the spring of 1940. He could not make up his mind. Was it right or wrong for science to serve war? His doubts were dispelled by President Roosevelt. Edward had never heard him speak. When it was announced that the President would address the Eighth Congress of Pan-American Scientists in Washington, Edward welcomed this opportunity.

". . . This very day, the 10th of May, 1940, three more independent nations have been cruelly invaded by force of arms . . . Belgium, the Netherlands, and Luxembourg," the President said.

"In this part of the world scientists are still able to pursue truth in freedom," Roosevelt went on to say. "But the distance from the battlefields is no safeguard to the scientists' freedom. . . . In modern times it is a shorter distance from Europe to San Francisco, California, than it was for the ships and the legions of Julius Caesar to move from Rome to Britain. Today it is four or five hours of travel from the continent of Africa to the continent of South America. . . ."

Instantly Teller arose to a lucid consciousness: there was an emergency beyond any doubt. The threat to the United States was real and great.

". . . You who are scientists may have been told that you are in part responsible for the débacle of today . . . but I assure you that it is not the scientists . . . who are responsible. . . ."

Now, Edward felt, President Roosevelt was addressing him; the President had sorted him out of the entire audience, had read his mind, and the words now pronounced were in answer to his doubts.

". . . The great achievements of science . . . are only instruments by which men try to do the things they most want to do. . . . Can we continue our peaceful construction. . . ? No, I think not. Surely it is time for our Republics . . . to use every knowledge, every science we possess. . . . You and I, in the long run, if it be necessary, will act together to protect and defend by every means at our command our science, our culture, our American freedom and our civilization."

Teller went into war work. To the uranium project he brought all the theoretical contributions that his rich mind could possibly yield.

He did not escape the fate of most physicists who were doomed to move over the country until the end of the war. He left Washington in 1940 to spend a year at Columbia University. From there he did not go back to Washington but moved on to Chicago, to California, and back to Chicago, according to the needs of the uranium project.

He was among the first to follow Oppie to Los Alamos. There I found him with his wife, his year-old son Paul, and the monumental grand piano that had followed him through his peregrinations. Like many theoreticians, Edward was extremely fond of music, and he devoted to it a good portion of his spare time. He practiced his piano a great deal, and, because he worked in the daytime, he played at night. This fact left his neighbors perplexed: should they be grateful to Teller for the beauty of the sounds reaching their ears, or should they deprecate the interrupted sleep?

Edward had become a prominent figure on the mesa by the time I arrived there. He was often seen walking absent-mindedly, with his heavy, uneven gait. His bushy eyebrows went up and down, as always when he was pursuing a new idea. He also helped his thought with unco-ordinated motions of his arms, and the leather patches on his elbows came in sight. It was smart to be thrifty in wartime. Theoretical men wore their sleeves out at the elbows, and their wives prolonged the suits' lives with leather patches.

Along with new ideas, Teller's mind produced doubts, scruples, uncertainties, changes of decision. His work, his duties toward it and toward his family, his responsibilities as a thinking citizen and as a scientist in wartime, all aroused questions in his mind, which mankind has not answered yet.

When he could forget his worries Edward delighted in simple pleasures. His favorite author was Lewis Carroll, and he started to read Carroll's stories and poems to his son Paul long before the child could understand them. Edward could be as playful and as naïve as his little boy, and each day the two of them spent some time entertaining each other. Edward started an alphabet for Paul, of which the following lines are an excerpt:

A stands for atom; it is so small
No one has ever seen it at all.

B stands for bomb; the bombs are much bigger,
So, brother, do not be too fast on the trigger.

S stands for secret; you can keep it forever
Provided there's no one abroad who is clever.

One of the best-known characters on the mesa was Mr. Nicholas Baker. In the Los Alamos array of faces wearing an expression of deep thought at all hours and under all circumstances, whether the men they belonged to were eating dinner or playing charades, Mr. Baker's face stood out as the most thoughtful, the one expressing the gravest meditations. He appeared to be dedicated to a life of the intellect alone, which allowed no time for earthly concerns.

When he walked about town, he did not seem to care where he was going. He let himself be led by his young son, a physicist like himself, who never left his side. Mr. Baker's eyes were restless and vague. When he talked, only a whisper came out of his mouth, as if vocal contacts with his fellow-men were of little consequence. He was a few years older than the other scientists—close to sixty in 1944—and all looked at him with reverence, whether they knew him personally or not.

Those who had known him in the past called him "Uncle Nick," for it was hard to say "Mr. Baker" but strictly forbidden to mention his true name. It was one of the best-guarded secrets that Niels Bohr was among us. Knowledge that an atomic physicist of such world-wide reputation was in Los Alamos would have been revealing.

It is surprising that the story of the vicissitudes that brought him from Copenhagen to Los Alamos in wartime were not made the subject of a novel. The warning he received from the Danish police that the Germans were looking for him; his escape from Denmark across the Sound to Sweden in a small craft; his flight to London under the auspices of the British government; his arrival in America with his physicist son, while his other sons and his wife remained in Sweden; his life as Mr. Baker; the fact that during the German occupation of Denmark his golden Nobel medal remained in Copenhagen right under the Nazis' eyes, dissolved in a bottle of nitric acid, and that it was recovered and recast after the war; all this constitutes dramatic material indeed.

Uncle Nick did not stay continuously in Los Alamos, but he spent most of his time there. He came often for a meal at the Peierls' in the apartment beneath ours. Genia Peierls was a woman of action. Wives of the British Mission were not given clearance to work in the Technical Area. They could not teach school because their background was too different from the American. Other possible jobs had little glamour: who cared, for instance, to help organize a mesa library so that Los Alamites would no longer have to rely for their reading material on the public library in Santa Fe?

The British wives would rather forego employment. Most of them were satisfied with their chores as housekeepers and mothers, with social activities, and with whatever else came up. Not Genia. Her exuberant constitution needed greater outlets, incessant action. She was always on the go. She made the rounds of the mesa, collecting bits of information, shedding advice. Early in the morning I often saw her leave her apartment, a knapsack on her back. I knew then that she was headed for the East Gate, where she would board the Army bus to Santa Fe. The bus was free to Los Alamos residents, and so were the jumpings, the swervings, the joltings that the springless vehicle, driven by a speed-loving GI, succeeded in achieving even on the smoothest stretches of road. Genia would come back in the evening under the load of a full knapsack, and on the following day there would be a dinner party at her home at which she would serve some low-point meat that she had been able to procure in Santa Fe.

To Genia a bachelor with a bad cold was a piece of luck. She could take his temperature, bring him fruit juices, nurse him in motherly fashion, whether he liked it or not. A man like Bohr, whose wife was on another continent, was almost as good. She could shed on him her warm affection, and feed him.

We always knew when Uncle Nick was at the Peierls'. Through the floor of our living-room characteristic sounds would reach our ears at characteristic intervals: loud and prolonged peals of laughter would alternate with perfect stillness. Uncle Nick's whispering voice as he told a joke to the Peierls did not carry upstairs. But a mere wooden floor could not dampen the sound Genia made when laughing. Bohr must have told many jokes, and all must have been funny.

Niels Bohr was still preoccupied about the fate of Europe. I remember a hike he and our family took together, during which he talked most of the time about the war, about Germany, and about the sufferings the Nazis had brought about. In Los Alamos, however, Bohr was less tense, less apprehensive, than he had been in 1939 in New York. The occupation of Denmark by the Germans, which Bohr had so dreaded in 1939, was a *fait accompli* after April, 1940. No event, frightening as it may be, is ever as frightening as the prospect of its happening. By now Bohr had accepted the status of things. His present grief was less poignant, less paralyzing, than the fears it had replaced.

On that Sunday in the fall of 1944 Bohr's mind could be diverted from his habitual worries now and then and made to focus on the marvels of nature that surrounded us. Because the air in the hills had already turned cold and the winds brought along an early chill in the afternoon, we sought the more sheltered path at the bottom of Frijoles Canyon, and we hiked down from the camping grounds at the Lodge to the confluence of the Frijoles stream with the Rio Grande.

We all stopped to observe the moves of a skunk, an animal whose strange habits are unknown in Europe. Its pretty appearance greatly delighted Uncle Nick. He squatted on his heels close to the little animal and excitedly praised it. He admired the fulness of its tail, the touches of white in its dark fur, the coquettish movements of its head. He was not aware of the dangers he was exposed to, and it took us a long time to persuade him to move away.

We went on, down the narrow gorge where the walls of the canyon are vertical and close together; where the stream jumps over huge boulders in tinkling cascades; where the Ponderosa pines grow taller than elsewhere, in their constant effort to find more light; where, when we stopped in stillness, we could hear the rattling noise of the rattlesnakes wiggling away in the sagebrush.

Bohr surprised us with his agility. He could be spry. We had to cross the stream numberless times, and he never stopped to consider its width, or the best place to go across it. He jumped it. And while he did so his body straightened, his eyes glowed with pleasure.

At the end of the Frijoles Canyon, where it opens into the valley of the Rio Grande, we stopped in silence. There is a sense of reverence in the perception of some landscapes. The river was full, thick, and red. The sand was white, with sparse plants of blooming cacti over it. In front of us stood a long, lofty wall, the end of which we could not see. It cut the pure blue sky sharply, and above it there was a small white cloud, all alone, soft, fluffy, bright with sunshine.

We climbed back, and Bohr was never out of breath. He maintained a good pace, and we could have gone no faster; at the same time he talked and talked about the war and Germany; but now and then his indistinct words were lost in the murmur of the stream. We let him do the talking and saved our wind for the ascent.

Bohr's youthful exploits did not amaze only the Fermis. On a bright Sunday a few months after this hike, when the snow had covered the Jémez hills, I was with my children on Sawyer Hill among a swarm of skiers from Los Alamos. Sawyer Hill was the best skiing slope within easy reach of our town. All went to ski there except for the few who, like Enrico, were sufficiently enterprising and energetic for cross-country excursions. Enrico, bored by the monotony of going up and down the same incline, after the first Sunday or two gathered a group that undertook greater feats, ventured on farther snow fields, climbed steeper mountains. In the evenings he was happy and proud if he could tell me he had out-tired much younger men than himself.

I never tackled more than the lowest part of Sawyer, and there I was on the Sunday when Bohr came with some friends and stood at the bottom of the slope. His eyes must have filled with nostalgic yearnings, with a strong desire for the sport he had practiced in past years, and a younger scientist offered to lend him his skis for a while. It was the younger scientist's undoing. Uncle Nick put the skis on and climbed away. He gave himself to elegant curves, to expert snow-plows, to dead stops at fast speeds, and to stylish jumps that no one else on the slope could perform. He went on with no pause for rest, with no thought for the man who had taken his place at the bottom of the hill, ski-less. He quit only when the sun went down and darkness and a chill descended upon the snow.

225

(22)

LIFE ON THE MESA

There are several ways of expressing the same concept. In his official report on atomic energy Mr. Smyth asserts that ". . . the end of 1944 found an extraordinary galaxy of scientific stars gathered on this New Mexican Mesa."

At about the time which Mr. Smyth refers to, General Groves summoned all Army officers stationed in Los Alamos and gave them a talk. The story goes that he opened his speech with the sentence: "At great expense we have gathered on this mesa the largest collection of crackpots ever seen." The "crackpots" were dear to the General, who went on recommending them to the good care of his officers.

A third way of stating the same idea is to say that Los Alamos was all one big family and all one big accent; that everybody in science was there, both from the United States and from almost all the European countries.

An intellectual *émigré* is a person selected by certain special traits of intelligence, initiative, adaptability, and spirit of adventure. Facts seem to prove that when these traits join those common to most scientists, they produce queer persons indeed. Hence General Groves's choice of the word "crackpots," which, we felt, applied especially to the numerous European-born men of science on the mesa.

"But I am an exception," Enrico said after relating to me General Groves's alleged speech. "I am perfectly normal."

We had just finished lunch, and Enrico was preparing to return to work. He rolled up his pants, straddled his bicycle, waved goodbye, and started up the steep street. In the effort to pedal uphill, he let the belt of his sport jacket ride halfway up his stooping back.

His shrunken blue-cloth hat, which he wore steadily both in rain and in shine, was perched precariously on top of his head. I wondered . . . normal . . . perfectly normal. . . .

Four minutes later I heard the one o'clock siren. At that precise moment Enrico would disembark from his bicycle in front of the Tech Area gate and would show his white badge to the guard. Enrico is never late, not even in the morning.

The first siren of the day went off at 7:00 A.M. It warned that work would start in an hour. Then Enrico stretched out in his bed, yawned, and remarked:

"Oppie has whistled. It is time to get up." Oppie was the director of the laboratories. If the sirens went off, it *must* be Oppie's doing.

In early morning there was a scramble in the house. The children were to get ready for school; Enrico took too long to shave in the bathroom which had no bath, but only a shower. Protests . . . shouts . . . a "Now it's my turn," in a shrill boyish voice . . . , an "I am older, I'll go first." An occasional fight. . . . Some unavoidable kicking under the breakfast table if the children sat across from each other; some work of the fists if they sat side by side.

Then the house was suddenly still. I did the dishes, started a soup that would be cooking all morning on the GI electric plate, and by nine I was at work in the Tech Area.

At that period wives were encouraged to work. There was great scarcity of clerical help at first, and some young men had been asked to join the Los Alamos group both because they were good physicists and because their wives were experienced secretaries. Harold Agnew, the student who had helped move the small pile from Columbia University, was of the number: Oppie had considered Beverley Agnew an additional asset and had hired both.

Apart from the shortage of woman power, which slowly decreased as single girls joined the project, it was an established policy to encourage wives to work. Colonel Stafford Warren, the head of the Health Division of the Manhattan District, placed little faith in women's moral fortitude. In the early days of the project he declared himself in favor of giving work to the wives to "keep them out of mischief."

The wives were only too happy for an opportunity to peek inside secret places, to share the war effort, to earn a bit of money. I worked three hours, six mornings a week, as clerical assistant to the doctor's secretary in the Tech Area. I was classified in the lowest category of employees, for I had no special experience or a college degree. When Enrico had asked me to marry him, I was halfway through school, and there seemed to be no point in waiting until I finished. Few married women had a career in Italy at that period, unless there was real need for more earnings in the family.

So in Los Alamos I was paid at the lowest rate for my three daily hours, which was not much; but I was kept busy, happy, and "out of mischief." I was given a blue badge that admitted me to the Tech Area but did not permit that I be told secrets; these were all saved for the white badges, the technical personnel. My boss, Dr. Louis Hempelmann, was very conscious of this fact the day he hired me. He was a tall, willowy young man with a shock of blond hair on his forehead. I was then thirty-seven years old and probably the oldest person he had ever employed. He was my first paying boss (I had done only volunteer work so far), and we both acted shy. His embarrassment showed in his easy blushing, which made him look little older than a schoolboy; mine in verbiage and too many questions.

He felt some explanations were due me about my future work and started by saying:

"Many of the boys are exposed to radiation: tube alloy. . . ."

"What's tube alloy?" I asked, puzzled by a term I had never heard before.

"Ask your husband," Hempelmann answered, blushing to the roots of his hair; then he went on:

". . . and 49. . . ."

"What's 49?" I asked again, for I had not yet gathered that my lack of understanding was due to ignorance not of chemistry but of the lingo developed for secrecy's sake.

"Ask your husband," Hempelmann repeated.

I knew better than to ask Enrico, because from him I could expect at most a noncommittal grin. To these questions, as to those roused by the work under the West Stands in Chicago, I found the answers in the Smyth report after the war had ended. Tube alloy was uranium and number 49 stood for plutonium.

My work kept me well informed about all sorts of inconsequential details. I knew who had a bad cold and whose splitting headache was relieved by aspirin from our office. Because I had been classified so low, my tasks were confined to preparing, filing, and bringing up to date personnel cards. I could also mark medical histories "secret" in red with a rubber stamp. I was acquainted with the number of corpuscles in many people's blood, and I learned immediately if a man had been transferred from one part of the project to another.

I passed on my information to Enrico, who never knew anything. He was associate director of the laboratories, but, to my great amusement, I was always the first to tell him the gossip of the Tech Area and the personnel movements.

Besides being associate director, Enrico was the leader of the "F Division," in which F stood for Fermi. When he arrived at Los Alamos, he managed to gather a group of very brilliant men. One of them was his imaginative friend Edward Teller; another was Herbert Anderson, Enrico's inseparable collaborator. No specific assignment was given to the F Division, but they solved a number of problems which did not fit in the work of any other division. It was typical of Enrico to be engrossed in his work and to pay no attention to what was going on around him.

At first sight, Los Alamos gave the impression of confusion; yet our life there was more than orderly, it was overregulated. Not only was our daily schedule adjusted to the sirens that announced beginning and termination of work, but we had also to submit to a number of rules set by the Army. In many ways we were a socialistic community that Army officials ran uncontested. Only the Town Council ever tried to oppose them. This body representing civilian residents loudly voiced the population's grudges and strove to teach democratic manners to the military rulers.

A large part of the administrative power rested with the housing office. For several months after its establishment this office was in civilian hands, and Rose Bethe was in charge of it until shortly before I arrived on the mesa. Rose, Hans Bethe's young wife, was born in Germany and educated in the United States at Smith College. This combination resulted in self-reliance, efficiency, and stubbornness; in firmness in the face of discontent and resistance to outside pressures. All these qualities were direly needed in the discharge of her duties.

Rose's first task was to assign lodgings to newcomers. She was instructed to distribute apartments according to the number of children, not according to personal wishes. Childless couples could have only a one-bedroom apartment; those with one child a two-bedroom one; those couples who, like us, had two children were given a three-bedroom apartment. Larger families were not considered. Rents were charged in proportion to the men's salaries, not to the number of rooms. Thus we paid over twice as much as our next-door neighbor, a machinist, for an identical apartment.

As could be expected, Rose displeased several tenants: she housed Edward Teller, who had no fixed hours to practice his piano, beneath a bookworm who needed tranquillity; the touchy Basilisk above a jazz leader, who assembled the full band in his home; a boy with a passion for chemistry and explosions in the same building as a brood of innocent children.

When Rose had taken office, several of the houses on "Bathtub Row" were already occupied. The amount of grumbling about the way she assigned the rest equaled the total grumbling on housing elsewhere, for the stake was much higher. "Bathtub Row" was the name that, in a stroke of wit, Alice Smith, a wife, had given to the small group of log and stone houses that had belonged to the Los Alamos school. They were attractive, well-built cottages, far more desirable than any later building. Besides, they were equipped with bathtubs, while Army-built apartments were provided with mere showers. This was one of the two main reasons why the homes in Bathtub Row were the most coveted on the mesa. The other was the social status that went with them.

In the beginning only the most important persons lived there: Oppie, the colonel, the Navy captain. So Bathtub Row acquired much glamour. As the months went by, it became uncertain in envious minds whether Bathtub Row derived its luster from its residents or whether the residents acquired distinction from living in it.

Whatever the case, women from shower-furnished apartments were convinced that the wives in Bathtub Row enjoyed too many privileges. The most unfair of these, it was whispered, was their getting as much help as they wanted, even after the maids were rationed. Like assigning homes, so distributing maids was the responsibility of the housing office. Early in the morning GI busses made the rounds of Spanish villages and Indian pueblos to gather the hired help for the site. The maids reported to the housing office, where they received their schedule of work for the day, usually two three-hour periods at two different homes.

Los Alamos spread out wealth for miles and miles around, like a volcano which suddenly ejects a lava of gold. A poor population, used to living on the scanty products of the soil, learned the feel of money. All men who were not in the Army, all women who could leave their babies, all girls who could take time off from school, came to work on the mesa. Household help was plentiful, in the beginning. As more Spanish and Indian women were employed in the Technical Area and as the Los Alamos population grew, help became inadequate. The WAC's at the housing office worked a complicated system of rationing. If I had not worked, I would have been entitled to no help at all, because I had no child under five years of age, no chronic illness, and no baby under way. My part-time work allowed me two half-days of help a week.

Enrico had his pet solution of the help problem. The times when servants were available in large numbers can be forgotten, he said. On the other hand, machines do only a small fraction of the housework. We may find other substitutes, besides machines, for maids and cleaning women. We may look outside the human race, and find it possible to train chimpanzees and gorillas to do menial tasks, to run a vacuum cleaner, to scrub floors, and wash walls. Perhaps they could learn to answer doorbells and to wait on tables. As Enrico saw it, the housing office in Los Alamos would set up an Agency of Primate Distribution to train and lodge primates. Housewives could hire them at low cost.

The trouble with Enrico is that he always keeps his bright ideas to himself. He did not try to sell this one to the housing office, and the maid shortage went on.

Another indication of the socialistic—or perhaps I should say "paternalistic"—trends in Los Alamos were the medical services, for which we did not pay. I always pitied our Army doctors for their thankless job. They had prepared for the emergencies of the bat-tlefields, and they were faced instead with a high-strung bunch of men, women, and children. High-strung because the altitude affected us, because our men worked long hours under unrelenting pres-sure; high-strung because we were too many of a kind, too close to one another, too unavoidable even during relaxation hours, and we were all crackpots; high-strung because we felt powerless under strange circumstances, irked by minor annoyances that we blamed on the Army and that drove us to unreasonable and pointless rebellion.

Our Army doctors were kept busy treating the minor ailments of a healthy population. They added wing upon wing to our hospital, which had looked like an H at first but had soon grown shapeless. In that hospital an unbelievable number of babies came to life at the standard price of fourteen dollars apiece, the cost of food for the mother. To the world at large all those babies were born in Post Office Box 1663 of Santa Fe.

The doctors' interest was sharpened and their professional skill challenged when an accident took place in the Tech Area. It was connected with secret work, and the circumstances under which it occurred were not divulged at the time. In 1952 a report of the accident was published by Drs. Hempelmann, Lisco, and Hoffman in the *Annals of Internal Medicine*.

It was the evening of August 21, 1945. After working hours two men went back to an isolated laboratory at the bottom of a near-by canyon. They resumed research on an experimental nuclear reactor called a "critical assembly," which they had pursued earlier in the day.

Nuclear reactors are arrangements in which a chain reaction may take place. An atomic pile is one. A chain reaction may occur in a reactor only when its dimensions have reached or passed the crit-ical size. A "critical assembly" reactor is ordinarily kept below its critical size so that it cannot chain-react. In the course of experi-ments its dimensions may be increased to produce a controlled chain reaction.

Of the two men in the laboratory on that particular evening, one had his hands on the reactor. His first name was Harry and he was twenty-six years old. The other man was at some distance.

Suddenly the reactor passed its critical size. A noncontrolled chain reaction started and delivered enormous amounts of radiation. Within twenty-five minutes both scientists were in the Los Alamos hospital and doctors were examining them.

Harry's hands were already badly swollen. The other man, who had been at some distance from the reactor, suffered no permanent injury.

Dr. Hempelmann was in charge of health in the Tech Area. In the days that followed, Harry's illness became the chief concern in Hempelmann's office, where I was working. The little information about Harry that was not withheld for security reasons was sufficient to arouse deep pity.

Harry was the first person in America to suffer from acute radiation. American physicians were at that moment observing the casualties of the atomic bomb at Hiroshima. There, however, the action of blast and heat had been mixed with that of radiation. Harry's case was unique. From the radioactivity found in samples of his blood it was possible to calculate the dose of radiation he had been exposed to. His right hand had received over two hundred thousand times the average daily dose to which men working with radioactive materials would normally be exposed.

I remember seeing the pictures of his hands at the office. They were taken at successive intervals and indicated with vividness, beyond any doubt, the rapidly increasing deterioration and the painfulness of his condition. Huge blisters, loss of skin, effects of poor circulation in his fingers, finally gangrene, showed in succession on the pictures. On the twenty-fourth day of his illness he died.

About a year later eight persons were involved in a similar accident, which also resulted in one fatality. By that time we had left Los Alamos.

While we were leading our secret life on the mesa, lulled in the illusion that nobody could detect us, the most fantastic speculations about the "hill" kept conversation lively in Santa Fe. Residents there could see our columns of smoke in the daytime and our lights at night. They were well aware that something tremendously hush-hush was going on at our place, the "hill" to them.

Santa Fe has preserved an ancient Spanish flavor in her architecture and in her habits. Quiet and peaceful, she invites rest. During the war, when pleasure-travel was reduced to a minimum, the town was dozing. Silent Indian women, bundled with their babies in their blankets, squatted on the floor of the roofed porch in the Governor's Palace among their pottery and their jewelry for sale. Dark-haired Spanish girls with bright-red lips roamed around the plaza in colorful dresses and threw shy glances at possible admirers. At all hours of the day Spanish-American men took their siestas on the benches in the plaza, shadowed by trees that had fostered the rest of generations. Through the doors of their stores salespeople peeked out or leisurely waited on customers who were never in a hurry.

Then came the women from the "hill" to upset the slow pace of the town. They poured out of overcrowded cars and scattered around with hurried, purposeful strides. They wagged their capacious shopping bags and filled them hastily to the brim. They had no time to spare. Work and children awaited them on the hill. They bought and bought. All goods that reached Santa Fe in those times of scarcity disappeared into the women's bags, from children's shoes to repair parts for washing machines.

Santa Féans did not know that the only stores on the hill were a commissary, where we bought our groceries, and a trading post that sold aspirin, pencils, and whatever else was on hand. Santa Féans speculated among themselves but asked no questions. They did not even ask where they should deliver parcels too heavy to be carried. They had learned that everything went to 109 East Palace. There, at the back of a Spanish patio, was the two-room "city office" of our project. Behind a desk an efficient woman, Dorothy McKibben, sat calm and unruffled, surrounded by large boxes and crates to be hauled by truck to the mesa and by the piles of smaller parcels that shopping women dumped on the floor to make room for further purchases in their bags. All women brought their difficulties and their checks to Dorothy. She indorsed the latter so they could be cashed at the bank, and smoothed out the first: Yes, she knew of a boys' camp; Yes, she could recommend a good eating place; Yes, she could arrange for a ride to the mesa later in the evening; Yes, she would try to get reservations at a good hotel in Albuquerque; Yes, she could give them the key to the ladies' room. Women always came out of Dorothy's office with greater cheer than when going in. They rushed out determinedly to accomplish more. Santa Féans looked on and asked no questions.

Then a day came in the month of December after the end of the war when the Army mitigated our seclusion and gave us permission to invite a selected and approved group of Santa Féans to Los Alamos. There was to be an official tour of the town and an official reception at Fuller Lodge, which had been the central building of the Los Alamos School for Boys and was now used as hotel and restaurant. Then the guests would go for dinner to the homes of various scientists.

That was the day of the Great Water Shortage.

Water had been scarce at all times, and the Army had issued frequent warnings: we should economize on showers, laundry, and dishwashing. As the number of inhabitants went up, the level in the reservoirs went down. By the summer of 1945 the shortage was acute. From our faucets came more algae and chlorine than water. We washed in cupfuls of this mixture and frowned upon residents of Bathtub Row when they indulged in a bath. It was not often.

Fate dropped her last straw on the eve of the day the Santa Féans were to come. During the unusually cold night a main froze, and by the morning nothing at all came from our faucets.

How would we wash the dishes after our dinner party?

At noon I was ready to start a stew for twenty, when a disheveled wife rushed into my apartment to say the party was called off. Fuller Lodge could not spare enough water to serve tea to the visitors.

The party was called on again just in time for my stew to get done. After a few hours of agitated consultations and contradictory conclusions, we heroically decided we would fill buckets at the trucks that had begun hauling water from the Rio Grande ten miles away. Dorothy McKibben had spent all day at her direct telephone line to Los Alamos and had transmitted our changing decisions to anxious Santa Féans. Finally she had handed them their passes.

The party was a success. Santa Féans at last filled their eyes with sights of the secret "hill." And we, the Los Alamos wives, felt the satisfaction of a hard-won victory.

Before effective measures were taken and a plentiful supply of water was assured to Los Alamos, the Great Water Shortage had disrupted life on the mesa. Mothers cried over the piles of diapers they could not wash; angry voices rose against the Army in the Town Council; a few people were scared away and left the project. Soon the Army organized a truck service to haul water twenty-four hours a day up the winding road from the Rio Grande Valley. For several months trucks brought up a hundred thousand gallons of water a day and slide rulers worked busily to figure out the cost of each gallon.

Before this happened, before we were allowed to receive guests in Los Alamos, the war had ended with the tremendous impact of two atomic bombs dropped on Japan.

(23)

THE WAR ENDS

Genia Peierls, who always managed to learn what was going on ahead of other wives, brought me the news on the morning of August 7. It could have been 10:30, and I was in my kitchen because I did not work during the children's summer vacations. I heard Genia rush upstairs. Her rapid steps imparted her inner agitation to the wooden staircase.

"Our stuff was dropped on Japan," she shouted upon reaching our landing. "Truman made announcement. It was transmitted ten minutes ago in Tech Area. Over paging system."

She stepped into my kitchen and stood, her brown eyes aglow, her large hands spread out, palms upward, her red lips parted.

"Our stuff." That is the word she used. Not even then, the morning after Hiroshima, had we, the wives, fully realized that Los Alamos was constructing atomic bombs.

Genia and I switched on the radio and listened. The secrecy of the bomb was finished, dead, to be forgotten at once.

"We repeat President Truman's words . . ." the announcer said, ". . . the first atomic bomb . . . equal to 20,000 tons of TNT. . . ."

How stupid of me not to have guessed! I had had my hints in the past. In 1939 I had heard that a chain reaction was theoretically possible. In 1941 a physicist's wife had given me the book by Nicolson that told of an imaginary diplomatic incident caused by the dropping of an atomic bomb. In 1943 Emilio Segré, on a visit to Chicago, had cheerfully greeted me with the cryptic assertion: "Don't be afraid of becoming a widow. If Enrico blows up, you'll blow up too."

Had I ever asked myself the reason for Emilio's words? Could they indicate dangerous work, which did not necessarily have anything to do with atomic explosions? Enrico worked under the West Stands of Stagg Field at that time, only three blocks from home. . . . Had I wilfully dismissed any hints I had received because I knew I could not ask questions and interest in Enrico's work was futile?

"It was to spare the Japanese people . . . that the ultimatum . . . was issued at Potsdam. . . . They may expect a rain of ruin from the skies, . . ." Truman's words went on in the announcer's voice.

I might have guessed, at least after the Trinity test, I thought to myself. On the other hand, I had heard very little about it. Early in July men had started to disappear from the mesa and the word "Trinity" had floated with insistence in the air. My boss, Dr. Hempelmann, had also gone to Trinity. By July 15 nobody who was anybody was left in Los Alamos, wives excepted, of course. On the afternoon of the fifteenth a woman physicist had told me that she, her husband, and some other young people would drive south to the Sandia Mountains near Albuquerque. They would climb on a peak and camp overnight. If they managed to stay awake, they might be able to see something of the test that would be carried on some hundred miles farther away.

On the next morning word spread through the usual grapevine that a sleepless patient at the hospital in Los Alamos had seen a strange light in the early morning hours. The test, it was thought, must have gone off successfully. Late that evening some of the men returned. They looked dried out, shrunken. They had baked in the roasting heat of the southern desert, and they were dead-tired.

Enrico was so sleepy he went to bed without a word. On the following morning all he had to say to the family was that for the first time in his life on coming back from Trinity he had felt it was not safe for him to drive. It had seemed to him as if the car were jumping from curve to curve, skipping the straight stretches in between. He had asked a friend to drive, despite his strong aversion to being driven.

A New Mexico paper mentioned the extraordinarily brilliant flash of light—perhaps a dump of ammunitions had blown up, it said. Even a blind girl had seen it.

I had heard no more about Trinity. Men had resumed their work at the usual fast pace.

"Mind me, Laura, at Trinity they must have exploded atomic bomb," Genia said, after we had listened to the radio. She was right.

On July 16 at Alamogordo (called "Trinity" for security reasons), in the southern part of New Mexico, the first atomic bomb ever made had been exploded. General Farrell, in a report to the War Department released to the press on the day after Hiroshima, used these words to describe the explosion:

"The whole country was lighted by a searching light with the intensity many times that of the midday sun. It was golden, purple, violet, gray, and blue. It lighted every peak, crevasse, and ridge of the near-by mountain range with a clarity and beauty that cannot be described. . . . Thirty seconds after the explosion came first the air blast, pressing hard against the people and things; to be followed almost immediately by the strong, sustained, awesome roar which warned of doomsday. . . ."

Now I could ask questions of Enrico. How would he describe the explosion? He would not be able to do it objectively, he said. He had seen the light, but he had not heard the sound.

"Not heard? How is it possible?" I asked bewildered.

All his attention, Enrico answered, was concentrated on dropping small pieces of paper. He watched them fall. As he had expected, when the air blast following the explosion hit them, it dragged them along. They fell to the ground some distance away from Enrico. He paced that distance counting his steps. He thus measured the path traveled by his bits of paper, and from it he was able to calculate the power of the explosion. His figures coincided with those of precision instruments and of accurate calculations. Enrico has always favored simple experiments. He was so profoundly and totally absorbed in his bits of paper that he was not aware of the tremendous noise, described by other witnesses as "a mighty thunder" and "the blast from thousands of blockbusters."

After he had completed his calculations, Enrico climbed on a Sherman tank lined with lead to screen the inside from radiation, and he explored the crater that the bomb had dug in the desert. A depressed area 400 yards in radius was glazed with a green, glass-like substance, the sand that had melted and then solidified again. It did not look at all like Meteor Crater.

Swift events followed Hiroshima. A second bomb was dropped on Nagasaki. Russia waged her six-day war on Japan. On August 14 Japan surrendered. As if caused by reverberation of the atomic bomb, an explosion of feelings and of words was set off in Los Alamos.

Women wanted to know. Everything. At once. But many things could not be said even then, cannot be said now. Children celebrated noisily, paraded through every single home, led by a band playing on pots and pans with lids and spoons. Men viewed the consequences of their work, and suddenly they became vocal.

Of the three groups' reaction, the men's, the women's, and the children's, the women's was the least peculiar. They behaved as all women would under the same circumstances. Their first bewilderment turned into immense pride in their husbands' achievement and, to a lesser degree, in their own share in the project. Los Alamos had caused the war to end abruptly, perhaps six months, perhaps a year, sooner than it would otherwise. Los Alamos had saved the lives of thousands of American soldiers. The whole world was hailing the great discovery that their husbands had given to America. The wives' elation was justified. When among the praising voices some arose that deprecated the bomb, and words like "barbarism," "horror," "the crime of Hiroshima," "the mass murder," were heard from several directions, the wives sobered. They wondered, they probed their consciences, but found no answer to their doubts.

Children of thinking age, like ours, became suddenly aware that, after all, their fathers, the men who did the things they were expected to do, who scolded and explained how to use a chemical set or how to solve a problem of geometry, who took too long shaving in the bathroom, who wore white badges on their coats, who ate their meals and went off to work, the men they were so used to they had never thought to evaluate, those men were important people. Perhaps even a wee bit more important than "So-and-So's father who is a captain in the Army" and in Giulio's mind was the most important man ever possible. Their fathers' names were now in black letters in the papers. They, the children, had lived in a place that made the headlines. Their school in Los Alamos had seemed small, not so well equipped as city schools. Nella had been upset, because high-school children of two different grades were packed into the same class. Now all this was reason for pride. The Los Alamos school was mentioned in newspapers, not the city schools. Nella's teachers had been no less than wives of great scientists, Mrs. Robert Wilson and Mrs. Cyril Smith. Our children and the Peierls' and the others in their age group went over and over these facts, and the feeling of their own importance grew in them.

Toward the end of August Giulio spent a week at a ranch for boys near Las Vegas, New Mexico. The owner of the ranch read an article in the papers in which Fermi's name was mentioned together with Einstein's.

"Are you any relation to this famous scientist, Fermi?" he asked Giulio.

"I am his son," Giulio answered, but the owner would not believe him. Giulio was a young boy like all the others, with no particular distinction about him.

I had similar experiences at various times. Once, while visiting Italy a few years later, I had a suit made by a modest tailor in a village in the Alps. He was a small man with a bad limp and witty eyes.

"Are you any relation to Fermi, the inventor?"

"I am his wife."

"Impossible!" was his immediate comment. What wife of a well-known man would go to him for a suit?

I was not prepared for the change that the explosion at Hiroshima brought about in our husbands at Los Alamos. I had never heard them mention the atomic bomb, and now they talked of nothing else. So far they had focused all their attention on their research, and now the entire world was their concern. To me they had seemed to be working with their usual zeal and dedication, and now they assumed for themselves the responsibility for Hiroshima and Nagasaki, for the evils that atomic power might cause anywhere, at any time.

Through the years that followed the opening of hostilities in Europe, the scientists in the United States had joined the war effort with remarkable readiness. A few, like Enrico, had had to make no transition: their peacetime research had suddenly revealed war potentialities, it *was* war work. A few had undergone a period of uncertainty, like Edward Teller. Once they made their decision, the scientists accepted it wholly, and never went back on it. War work became their normal work, and into it they brought their working habits.

Scientists have always lived in a certain protective isolation from the rest of the world, within the walls of their proverbial ivory tower. They were not concerned with the practical use of their achievements. Inside the ivory tower, contributions to science were an end in themselves.

Enrico is fond of stressing this point. It used to be one of his favorite topics in popular lectures. When he was very young and could not improvise his speeches, as he does now, he used to dictate them to me. Many started with such words as: "When Volta in the quiet of his little laboratory. . . ." The gist of the story was that when the famous Italian physicist discovered his voltaic cell he was living in his ivory tower. Neither he nor his contemporaries had foreseen the consequences of his work. Electricity, studied by a handful of researchers, was confined to laboratory experiments. Half a century was to pass before it became the dominant factor in the marvelous inventions that have revolutionized our mode of life.

Our husbands were not different from other generations of scientists. Helped by the physical separation of Los Alamos from the world, they worked in a certain isolation. They knew they were striving to make something that would likely shorten the duration of the war. It was their duty to concentrate all their powers upon this single aim.

Perhaps they were not emotionally prepared for the absence of a time interval between scientific completion and the actual use of their discovery. I don't believe they had visualized a destruction whose equivalent in tons of TNT they had calculated with utmost accuracy.

Other groups of scientists elsewhere had given more thought than the men in Los Alamos to the intricate problems that would ensue from use of the atomic bomb.

Pressure of work at the Chicago Met. Lab. was somewhat relieved when the phase of production was attained. Then other projects—Hanford, Oak Ridge, and Los Alamos—had taken over various activities, and the greater pressure was on them. The men at the

General Groves Pins the Medal of Merit on Fermi: The Other Scientists
Who Received the Medal Are (*left to right*) Harold Urey,
Samuel Allison, Cyril Smith, Robert S. Stone

Edward Teller and Fermi

The Institutes for Basic Research Are across the Street from ....

Better Homes and Gardens

. . . the Old West Stands of Stagg Field

About 1930: A Laboratory in the Physics Building in Rome

Twenty Years Later: The Giant Cyclotron in Chicago
(John Marshall Is Pleased with It)

Stephen Lewellyn

Met. Lab. could divert part of their energies and time to envisage and analyze the probable consequences of an atomic explosion.

A man of such a fertile imagination as Leo Szilard could not fail to foresee the difficulties that atomic power would bring in international relations. In March, 1945, he wrote an extensive memorandum in which he advocated international control of atomic energy and gave suggestions for a study of the modalities to be followed in its application. President Roosevelt, to whom the memorandum was addressed, died before he could see it. Szilard presented it to James F. Byrnes on May 28.

Three weeks earlier the war in Europe had ended. Germany had surrendered unconditionally on May 7. The fear that Germany might make atomic weapons and use them against us was gone. Left alone, Japan had no chance of victory. It would be defeated in the long run. The scientists at the Metallurgical Laboratory asked themselves whether there was any point in using the atomic bomb at all.

A committee on "Social and Political Implications of Atomic Energy" was appointed by the director of the Met. Lab. The committee included seven men, and Professor James Franck was chosen among them to be the chairman. On June 11, 1945, they submitted a report to Secretary of War Henry L. Stimson. In this report the seven scientists not only advocated international control but took a firm stand on the use of an atomic bomb: if we Americans should explode such a destructive bomb on Japan, we would place ourselves in a poor position to propose ban of atomic weapons and international control, once the war was ended. The committee recommended a technical demonstration of the new weapon in front of representatives of the UN. Recommendations along these lines were reiterated by a group of 64 scientists connected with the Metallurgical Project, in a direct petition to President Truman.

These views were not unanimous. Scientists agreed that use of atomic bombs would likely shorten the war and save lives, both American and Japanese. Opinions differed in evaluating this advantage against the danger of jeopardizing the future of international control, of world government, of everlasting peace.

A poll taken by Compton among more than 150 scientists revealed that, while many were in favor of a "preliminary demonstration on a [strictly] military objective," still opinions varied from use of the bomb "as the Army may see fit" to "keeping the existence of the bomb a secret."

The final decision rested with the president and with the strategists. Secretary Stimson's memoirs later revealed a few of the steps that led to the decision. On Stimson's advice President Truman appointed an "Interim Committee" of military and laymen best qualified to recommend war and postwar policies on atomic energy. The committee was assisted by a panel of four scientists: Compton, Lawrence, Oppenheimer, and Fermi. Secretary Stimson and the Interim Committee independently reached the same conclusion. The bomb was used.

In Los Alamos the paging system announced the news in the Tech Area, and the men were stunned. A blow is no less painful for being expected.

As the papers published descriptions of destruction in Hiroshima in greater and greater detail, the men in Los Alamos asked themselves whether they could truly delegate all moral responsibility to the government and to the Army.

To moral questions there are no universal answers. The range of reactions among our husbands was wide. Some felt that a rapidly ending war more than compensated for the destruction at Hiroshima and Nagasaki. Some told themselves that evil lies in the will to wage wars, not in discovery of new weapons. Some men said the atomic bomb should never have been built; researchers should have stopped working when they had realized that the bomb was feasible. Enrico did not think this would have been a sensible solution. It is no good trying to stop knowledge from going forward. Whatever Nature has in store for mankind, unpleasant as it may be, men must accept, for ignorance is never better than knowledge. Besides, if they had not built an atomic bomb, if they had destroyed all the data they had found and collected, others would come in the near future who in their quest for truth would proceed on the same path and rediscover what had been obliterated. Then in whose hands would the atomic bomb be placed? Worse evils could be conceived than giving it to the Americans.

Other men would have liked to hide or run away. Accusing voices rose from many parts of the world and further stirred the consciences. In Catholic Italy the pope had shown disapproval of the new weapon, and the Italians were uncertain how to judge it. Enrico received a letter from his sister Maria, which said:

"Everybody is talking about the atomic bomb, of course! Everyone wants to have his say, and we hear the biggest nonsense. People of good judgment abstain from any technical comment, and realize that it would be vain to seek who is the first author in a work which is the result of a vast collaboration. All, however, are perplexed and appalled by its dreadful effects, and with time the bewilderment increases rather than diminishes. For my part I recommended you to God, Who alone can judge you morally."

Among the scientists in Los Alamos the sense of guilt may have been felt more or less deeply, more or less consciously. It was there, undeniably. But it did not cause demoralization, it caused hope.

These atomic bombs are too destructive, men said in Los Alamos after Hiroshima. They will not be used again. There will be no more wars. The atomic era must and will be an era of international co-operation, in sharing the benefits of atomic power for peaceful uses, in banning atomic weapons. There must be an international controlling agency and a system of inspection of atomic research and of atomic industry. Such a system presupposes mutual trust between the nations, and mutual trust will bring world government. Once this is attained, there will be no more fear of war. Everlasting peace, the dream of sociologists and pacifists, will have come true.

What country, they went on arguing, would refuse to go along with such a program? What country could prefer utter destruction to survival? We, the most civilized nation, shall take the lead, show good will and confidence in the others; then they will follow us, they will open their doors to the international controlling body, they will relinquish their sovereignty to world government.

In October, 1945, scientists who thought along these lines formed the Association of Los Alamos Scientists (which on the following January merged with other similar groups into the Federation of American Scientists). Their central policy, stated in a newsletter, was to "urge and in every way sponsor the initiation of international discussion leading to a world authority in which would be vested the control of nuclear energy."

Prompted by a crusading spirit, members of the Association of Los Alamos Scientists sought opportunities to bring their views to the public, to promote public understanding and free exchange of ideas between themselves and laymen. They drafted statements, they wrote articles, they gave speeches.

Enrico did not share many of these views. He used to say that historical precedent, for what it is worth, does not show that improvement of weapons frightens men into not waging war. He also thought that the harshness of a war is not so much determined by the technical advance of the means of destruction but is rather controlled by the will to use the weapons and by the amount of punishment the fighting countries are willing to take. Enrico did not think that in 1945 mankind was ripe for world government. For these reasons he did not join the Association of Los Alamos Scientists.

The exodus from Los Alamos started toward the end of 1945. Enrico and many others with him felt that our country needed new generations of scientists as well as better weapons. Four years of war and of war work had kept young men away from the universities, and it was high time that scientists undertook to refill their depleted ranks. Besides, in peacetime many preferred to teach and to do research in fields untrammeled by secrecy rather than pursue a work that they had carried on with enthusiasm while our country was threatened. And so we left.

We took along with us whatever souvenirs we possibly could: Indian pottery and jewelry, cacti and pictures. But Herbert Anderson was more successful than anyone else in his choice of a souvenir. While he was living in Los Alamos, Herbert had purchased a horse, which he loved. Rather than leave him behind, Herbert ordered a special trailer, coaxed the unwilling horse to climb into it, and departed. Herbert took the poor animal in tow the thirteen hundred miles to Chicago. There he made passers-by turn around when he rode his horse on Hyde Park Boulevard, hitched him to a garden fence, and went inside to pay a call on his friends.

His friends were the Fermis. We had left Los Alamos for Chicago on New Year's Eve, half an hour before the beginning of 1946. Thus we brought to a close one of the most memorable periods of our life.

We were not the only persons who left Los Alamos with regret.

After the years spent together, after the sharing of a single purpose, it was sad for all to separate and to scatter over the country.

Our husbands had enjoyed the work in common, the co-operation between the various branches of science that are usually confined in university departments. To protract this co-operation in peace-time, to keep alive the spirit of Los Alamos, several of our friends joined the Institutes for Basic Research at the University of Chicago.

The idea of a research institute was born in Chicago during the spring preceding the end of the war. Arthur Compton had pondered for a while how best to keep together some of the physicists, biologists, chemists, engineers, and even metallurgists, whom he had first gathered at the Metallurgical Laboratory. He talked to Mr. Hutchins, then president of the University of Chicago, a man who always welcomed new and imaginative proposals. Several possible members of the future institute were approached by correspondence, and the plan took shape.

By the middle of July it was felt that correspondence was no longer adequate, that a meeting was necessary between representatives of the university and a few scientists. Harold Urey, Samuel K. Allison, Cyril S. Smith, and Fermi ought to be consulted. But the last three were extremely busy in Los Alamos during that July of 1945. They could not go to Chicago. Vice-president of the university Gustafson, Walter Bartky, dean of the division of the physical sciences, and Harold Urey were willing to take a trip to New Mexico, but they had no pass to Los Alamos. The six men met in Santa Fe, on the terrace of Dorothy McKibben's home, on top of a hill with a view over the golden vastness of the desert between the town and the distant mountains.

Over a lunch of sandwiches packed at Fuller Lodge on the mesa the policies for the future institute were discussed. It would not be divided into departments. It would provide a meeting ground for science and industry. The industry might give financial support to the institute and in return receive scientific advice and information on progress of research.

The new institute needed a director, and the six men consulted with one another. Harold Urey said he had tried administrative work, and he felt he was not suited for it. Fermi had never done administrative work, but he was sure he was not suited. Cyril Smith was a metallurgist who had been in industry before joining the uranium project. He had no experience in university work, he said. Sam Allison could not think of a good excuse and was named director on the spot. Longer and more protracted considerations could not have resulted in a better choice. Yet Sam Allison voiced some doubts: the responsibility of directing research in biology and metallurgy, besides that in physics and chemistry, was too great. Biology and metallurgy were remote from his field of research. Three Institutes for Basic Research were established in the end: the Institute for Nuclear Studies, the Institute of Metals, and the Institute of Radiobiology. Allison remained the director of the first.

As soon as the meeting was over, that same afternoon, the three men who had come from Chicago took the train back home. In Chicago, upon their return, they were told that only a few days previously the first atomic bomb had been exploded at Trinity, that the three men who had gone from Los Alamos to the meeting in Santa Fe had taken an active and greatly important part in the test. To the present day Dean Bartky has not recovered from the surprise he experienced then: Allison, Smith, and Fermi had seemed so calm, so matter of fact, so utterly like themselves at less exciting times!

The Institutes for Basic Research started to function at the beginning of 1946. And so we came to live in Chicago.

In Chicago on March 19, 1946, Enrico and four other scientists received the Congressional Medal for Merit for their help in developing the atomic bomb. The medal was awarded by the President of the United States "in accordance with the order issued by General George Washington at Headquarters, Newburgh, New York, on August 7, 1782, and pursuant to act of Congress."

Major General Leslie R. Groves, head of the Manhattan District, presented the medals to Harold C. Urey, Samuel K. Allison, Cyril S. Smith, Robert S. Stone, and Fermi in a simple ceremony at the Oriental Institute of the University of Chicago.

The citation accompanying Enrico's medal reads:

Dr. Enrico Fermi for exceptionally meritorious conduct in the performance of outstanding service to the War Department, in accomplishments involving great responsibility and scientific distinction in connection with the development of the greatest military weapon of all time, the atomic bomb. As the pioneer who was the first man in all the world to achieve nuclear chain reaction, and as Associate Director of the Los Alamos Laboratory, Manhattan Engineer District, Army Service Forces, his essential experimental work and consulting service involved great responsibility and scientific distinction. A great experimental physicist, Dr. Fermi's sound scientific judgement, his initiative and resourcefulness, and his unswerving devotion to duty have contributed vitally to the success of the Atomic Bomb project.

(24)

EXIT PONTECORVO

In the development of atomic energy the Manhattan District had used that accidentally discovered process for which Fermi and his collaborators had filed a patent application in Rome on October 26, 1934: in all atomic piles the neutrons emitted in uranium fission are slowed down by the layers of carbon before they bombard other uranium and transform it into radioactive plutonium. A major practical application of their patent had seemed very improbable to the group of Roman physicists in 1934. It had come true eight years later, a short span for the results of research to be utilized by technology.

It will be remembered that the conclusive experiment on the action of slow neutrons was performed in the goldfish fountain at the back of the old physics building in Rome—those goldfish have a right to be numbered among the forefathers of the atomic age. Senator Corbino had urged his "boys" to take a patent on their discovery, and, after their first bewilderment at the novel idea, the "boys" had followed his advice.

The group seeking an Italian patent was made up of seven men, five of whom claimed they were the actual inventors of the slow-neutron process: Fermi and his collaborators: Rasetti, Segré, Amaldi, and Pontecorvo. Of the remaining two, Professor Trabacchi, the "Divine Providence," had furnished the radon used in the research; D'Agostino had helped with the chemistry of the experiments.

An Italian patent was granted on February 2, 1935. It was only natural that the group should try to obtain patents in other countries. They did not know how to go about this; they were not businessmen and had no business connections abroad; they did not want to take time from research to deal with lawyers. A lucky circumstance aided them.

Among Fermi's and Rasetti's early students, together with Segrè and Amaldi, there had been another young man, Gabriello Giannini. Giannini had taken up physics with the intention of entering industry. After receiving his degree in physics, he had worked a while with a radio firm. But working opportunities in Italy held little promise to Giannini, who hoped for a rapid and successful career. He wanted to get married, to raise a family, and to gain a substantial share of what this world can offer.

Giannini went to seek his fortune in the United States in 1930.

Gabriello had the stuff of the executive: he talked easily and knew how to bring his qualities into light. He had faith in himself, stubbornness in the face of difficulty, enterprise, and the will to succeed. Despite the depression of the early thirties, he soon managed to set himself up in business.

To his old teachers and fellow-students he seemed the very man to help them get foreign patents. They made a deal. Giannini would assume the task of obtaining and exploiting patents in Europe and in America, and he would become the eighth man to share the not very probable proceeds.

The patents Giannini obtained in European countries are of no interest. In the United States G. M. Giannini & Co., Inc., filed an application for a patent with the Patent Office in October, 1935. The patent was to be taken in the name of the five inventors: Trabacchi, D'Agostino, and Giannini were shareholders by written agreement.

For a long time nothing happened. Americans must be of less trusting temperament than Italians. The Patent Office was cautious; almost five years went by before it became satisfied that the applicants were the actual inventors of the slow-neutron process. The American patent was granted on July 2, 1940.

Meanwhile, the historical events of the late thirties, the suddenly overpowering influence of naziism over fascism, had prompted Fermi and three more inventors to leave Italy.

Segrè had settled in Berkeley, California, with his family— he later joined the Los Alamos project, where he and Fermi found themselves working together after almost a decade.

Rasetti, who was still a bachelor at that time—he married in 1949 when forty-eight years old—joined the faculty of Laval University in Quebec and went there with his mother in July, 1939. Quebec was a quiet place to live in, and Canada a land abounding in trilobites. Rasetti pursued his activities as a physicist in Quebec, and at the same time he became a prominent geologist.

The youngest of the five inventors, Bruno Pontecorvo, "the cub," had gone to Paris in 1936 to study with Joliot-Curie. He decided not to return to Italy. In Paris he married a young Swedish girl, Marianne Nordblom.

Amaldi alone remained in Italy.

All sorts of difficulties hampered a settlement of patent rights. The secrecy of the uranium project during the war—lawyers of the Manhattan District could have dealt with Fermi or with Segré, not with Giannini—the change of management from the military Manhattan District to the civilian Atomic Energy Commission; the passage of the Atomic Energy Act of 1946, which provided for "just compensation" of patents used, but which delayed action, as all new legislation is likely to do; and the fact that Enrico became a member of the General Advisory Committee to the Atomic Energy Commission. For this post Enrico received no salary. Still he was a government employee, government lawyers said, and, accordingly, neither he nor the other inventors could place a claim against the United States.

When Enrico's term on the General Advisory Committee ended, Giannini resumed his efforts to reach an agreement with the Atomic Energy Commission. In a surprise move, without consulting the inventors, the G. M. Giannini corporation filed a suit against the United States government in the United States Court of Claims, on August 21, 1950. The suit charged the government with using the patented process without compensation. The Giannini corporation asked $10,000,000. The figure seemed shocking and preposterous to the inventors, but, Giannini explained, the sum claimed in a suit has little to do with the sum it is hoped to get. A figure must be stated as a starting point in any discussion.

Then, on October 21, before any action on the suit had started, the newspapers published the unbelievable story that one of the five inventors, Bruno Pontecorvo, had disappeared, vanished, possibly behind the Iron Curtain.

Giannini was perturbed. He did not at all like to represent a man who was probably a Communist, and who had likely fled to Russia, in a suit against the United States government. The G. M. Giannini corporation withdrew the suit.

Later, Giannini resumed dealings with the Atomic Energy Commission for an agreement on patent rights. A settlement was reached in the summer of 1953 for a fraction only, not of the amount Giannini had asked in his suit, but even of that he had hoped to obtain. Besides, the expenses had run so high that each inventor and shareholder received one-tenth of the sum obtained rather than one-eighth.

It seemed fantastic, inconceivable, that Pontecorvo might have fled to Russia with a wife and three children. The British atomic scientist Bruno Pontecorvo, the papers said on October 21, was believed to have slipped behind the Iron Curtain because he had the international police on his tail. Pontecorvo had worked at Harwell until the previous July; then he had gone to Italy for a vacation with his family. There had been no interest in his whereabouts until the British Intelligence had questioned the Italian Security Police. It was then ascertained that the Pontecorvo family had left Rome on a Swedish airliner, possibly direct to Poland. (Later news put Poland out of the picture and established that the Pontecorvos had left on September 1.) One story, confused though it was, gave detailed and accurate information of his stay at the University of Rome in the early 1930's; and it related that his friends still remembered him affectionately as the "kid." Pontecorvo's father in Milan had denied any knowledge of his flight, the story went on, and he had expressed the opinion that Bruno would go back to England when his time was due.

At first, Enrico and I were inclined to share the views of Bruno's father. We knew that several members of the Pontecorvo clan were prominent Communists in Italy. Still we felt Bruno was too level-headed to take such a drastic and irrevocable step as a flight to a Soviet country. He was perhaps skiing in some remote part of

Scandinavia, and he would reappear, once he became aware of the fuss he was causing.

Together Enrico and I went over all we knew about the cub after he had left Rome. Bruno Pontecorvo had lived in Paris until the Nazis had entered it. He had reached the south of France on his bicycle, while his young wife and their baby boy had been able to travel by train. Together, the family had crossed from France into Spain, then into Portugal. They had boarded a boat to the United States and had landed in New York on August 20, 1940. Two days later Bruno had his twenty-seventh birthday.

We saw Pontecorvo several times between his arrival in New York and his disappearance. We met his wife once. We never saw the boy who was born in Paris, nor the two younger sons, born later in Canada.

Bruno came to visit us in Leonia, shortly after he had arrived in this country. He was alone. Marianne, he said, was worn out by the sea-crossing and needed rest. There was nothing strange in this fact: an ocean journey on a boat overcrowded with European refugees could be no pleasure trip. We thought it somewhat peculiar that Bruno should refuse my offer to go see his wife, that there should be absolutely nothing I could do for them, that he should not introduce little Gil to us.

Bruno had not changed at all. Handsome, pleasant, showing little worry for his future, he was still the cub we had known in Rome. He had no position waiting for him in America, no definite plans. Nonetheless, he was outwardly serene, and he talked lightly of his predicament.

Perhaps a man of Pontecorvo's intelligence and likable disposition, with his well-established reputation of being an indefatigable worker, had a right to be an optimist. Soon he became scientific researcher for an oil company in Oklahoma. He was invited to join the Anglo-Canadian uranium project in early 1943.

He came to see us in Chicago on his way to Canada. His wife and son were not with him. He came again in 1944 before we left for Los Alamos. He had broken a leg while skiing in the Canadian mountains, but he did not seem to mind it. He hopped briskly, with elegant motions, even though he bent on crutches, and he enjoyed being the center of attention. He smiled cheerfully at anxiously inquiring friends, and then he hopped more briskly. He was in Chicago on a mission, sent by the Anglo-Canadian uranium project.

Bruno stayed in Canada six years. Early in 1949 he joined the British atomic project at Harwell. Toward the end of 1948 Ponte-corvo took his last trip to the United States, and this time his wife came along. Both of them came to see us one evening, as they were going through Chicago. They were late for dinner, a fact that upset Bruno out of proportion to the trouble it might have caused. Mari-anne had gone shopping and had lost her way back to the hotel, he said. He had scolded her; she should have known better than to be late. He had called a taxi, but he could do nothing more. As Bruno talked, as easily and volubly as ever, with unmistakable annoyance in his voice, his wife kept silent.

She was small and fair. She looked extremely young; it seemed impossible that she had given birth to three children. She sat on the edge of her chair. And she was painfully shy. All my attempts at making friends with her broke against the iceberg of her shyness. She thawed one moment only, when after dinner she watched me load my automatic dishwasher. Flickers of interest enlivened her baby-blue eyes.

It was surprising to hear later, after the Pontecorvo family had vanished, that Marianne was a Communist, like Bruno. British au-thorities received a report from Sweden to that effect. It was even more surprising to read between the lines in some stories about the Pontecorvos that she might have played a more important role than was suspected.

When I became aware of this hint, there was no longer doubt that Bruno and his family had slipped behind the Iron Curtain. All indi-cations pointed that way: the family's moves had been retraced from Rome to Copenhagen on a Swedish airliner; from Copenhagen to Stockholm by train; from Stockholm to Helsinki in Finland on a plane of the Scandinavian Airlines. It was reported that on the flight to Copenhagen Bruno was carrying a bulging briefcase and that he insisted on keeping twenty pounds of hand luggage by his seat. In Stockholm the Pontecorvos had not visited Marianne's parents, whose home was fifteen minutes by streetcar from the station; nor had the parents been called on the telephone. It was also reported that Bruno's youngest son, five-year-old Antonio, had told a fellow-passenger, "We are going to Russia."

The Pontecorvos arrived in Helsinki on September 2. No record of them was found in Finland on the list of tourists or among the passengers flying back to Stockholm. All other routes—air, boat, train—led to Russia. An Italian newsman flew to Helsinki to look for further clues. He reported in the Italian paper *Il Tempo* that the Pontecorvos had been met at the Helsinki airport by a car of the Soviet Legation, which took them to the harbor; that in the harbor the ship "Bellostov," scheduled to sail at 10:40 A.M., had waited until 5:00 P.M. and that it had weighed anchor as soon as the Pontecorvos arrived.

On November 6 Mr. Strauss, British minister of supply, speaking about Pontecorvo in the House of Commons stated: "While I have no conclusive evidence of his present whereabouts, I have no doubts that he is in Russia."

Enrico and I came to accept the fact that Bruno and his family had passed to the other side of the Iron Curtain.

Two main questions about the cub arose in our minds: What did we know of his political ideas? Could he have passed information to Russia?

We came to the incredible realization that we had never talked of politics with Pontecorvo. We knew the political opinions of all our other friends, and we could predict their stand on most issues. Not of Pontecorvo. To us he was the pleasant cub, fond of sports and games, who had never outgrown the attitude of the college boy on an athletic team. To us he was the man we had last seen in 1949 in Basel and in Como, at two consecutive meetings of physicists. In Basel he and Enrico swam a mile in the Rhine. In Como the two of them played tennis together. And these are our last recollections of the cub. Politics, debates on the Communist and capitalistic systems, have no place in our memory.

The second question was as hard to answer as the first. So far as Enrico knew, Pontecorvo had never had access to vital information. At Harwell he was doing research on cosmic rays, not on secret work. The British minister of supply in a debate in the House of

Commons stated on October 23: "For several years Dr. Pontecorvo's contacts with secret work have been very limited."

If Bruno had been linked with the Fuchs case, as some newspapers suggested, then why did he wait so long to make his escape, from March, when Fuchs was sentenced, to September? At Harwell he was well liked, he had many friends. They had noticed nothing unusual in his conduct. The Italian friends and relatives who saw him during the summer of 1950 say that he was his usual self, carefree and enjoying his vacation, at least until late in August.

On the other side of the picture are these facts: that shortly after the Fuchs trial Pontecorvo had gone of his own accord to the British Intelligence and had volunteered the information that one of his brothers was a Communist. That in the summer he had resigned from his post at Harwell, after a long period of indecision, and had accepted a position at the University of Liverpool. (He was to start on his new job in January, 1951.) That on August 22, nine days before he left Rome, Bruno met his Communist brother and sister-in-law and that his mood seemed to change after the meeting.

Over three years have now passed since the Pontecorvos' disappearance. No word has been heard from them. Nobody has seen them. Their relatives deny knowing anything at all about them. The British government has made no charge against Bruno. If anything at all has been found in England that could be construed as evidence against him, the existence of this evidence has never been revealed. And all this happened in the twentieth century!

(25)

A NEW TOY: THE GIANT CYCLOTRON

Cyclotrons, like the Pyramids of Egypt, may go down in history as examples of nonutilitarian monuments," Enrico said. Almost nine years had passed from the memorable day when physicists had first operated a pile under the West Stands of Stagg Field. Across the street from the make-believe medieval castle a group of new buildings had sprung up, as modern, as neat and elegant, as the castle was shabby and decrepit-looking: the Institutes for Basic Research and the Accelerator Building.

The giant cyclotron of the University of Chicago was erected inside the low, elongated Accelerator Building. As Enrico was talking to me about it, the cyclotron had just started operating, four years after its construction was undertaken in July, 1947. Enrico was as excited and pleased as a child who has received a new toy long dreamed of and exceeding expectations. He played with the cyclotron at all hours of day and evening during that summer of 1951. He allowed the cyclotron to upset his routine.

"Cyclotrons and pyramids," I replied. "What an odd comparison! What do they have in common?"

"They are both tangible victories of men over the brute powers of matter. They were both built with no consideration of financial returns."

These assertions were true. Pyramids were meant to flatter royal vanity during kings' lives and to protect royal bodies after kings' deaths. Cyclotrons may advance man one small step on the path of knowledge. Cyclotrons and pyramids are nonutilitarian monuments indeed!

A hundred thousand slaves worked on the Great Pyramid of

Cheops. With ropes bound around their bare shoulders, they dragged stones of two and a half tons each on sledges and rollers from the quarries in near-by mountains to the site of construction.

The erection of giant cyclotrons five thousand years later, in the age of machines, was not so dramatic a feat, but an achievement of great technical skill. Ready-made cyclotrons for sale do not exist, and this is another feature that cyclotrons share with pyramids.

The first giant cyclotron was built by Ernest Lawrence, the father of all cyclotrons, on a hill near Berkeley in California, immediately after the second World War. It was safer to stay away from habitations, Lawrence felt, because such a big cyclotron would send out deadly radiations while operating. Other universities also planned to build cyclotrons away from campuses.

But the Chicago physicists are lazy. They wanted their cyclotron at hand, right where they did the rest of their work and their teaching. Full protection could be given both to workers and to persons living in the neighborhood. The cyclotron would be erected in a pit, deep below street level, and most of the radiation would be absorbed by the ground. For further protection the entire cyclotron would be inclosed in a thick shield of reinforced concrete.

The plans for the Accelerator Building were drawn according to this project: the building provided space for other types of accelerating machines, a pit for the erection of the cyclotron, and a long bay, in which a huge crane, capable of lifting a hundred tons at one time, can be brought where it is needed.

It was hoped that one of the big industries would build the cyclotron for the University of Chicago. But scientists and industries could not reach an agreement on this project. The price requested seemed too high, parleys dragged on, and time went by. Herbert Anderson, who is of impulsive nature and likes action to follow right after ideas, decided he would build the cyclotron himself. What did industrialists know about cyclotrons, anyhow? Herbert asked. He would have to give them all the specifications and to supervise the work in any case. He could just as well do the construction. Anderson was assisted and helped by another physicist, John Marshall. Liberal financial support was received from the United States Navy, through the Office of Naval Research. Part of the expense was met by Chicago citizens. The Chicago cyclotron cost two and a half million dollars, little more than the amount spent on feeding the workers on the Cheops pyramid.

The essential parts of a cyclotron are a huge magnet and a metal box. The metal box of the Chicago cyclotron is so large that it could be used to store three hundred bushels of grain. However, all that space is wasted: the box is kept empty. Indeed, it is more than empty: a vacuum is made in it by means of nine big vacuum pumps. The particles to be accelerated are sent inside this box. The magnet bends their paths and keeps them inside the box, while a radio frequency field makes them run faster and faster.

The magnet of the Chicago cyclotron is made of a steel core and of copper coils wound around it. When an electric current is sent through the coils, the steel core becomes magnetized. The magnet weighs 2,200 tons, almost forty times as much as the magnet of the cyclotron which Lawrence built in the thirties.

The sections of the core were forged by Bethlehem Steel in Pennsylvania. They weigh as much as eighty tons each. A special crane had to be used at the railroad station in Chicago to transfer them from the train to tractor-trucks, built especially for this purpose. Once the trucks were loaded, they were routed over streetcar rails wherever it was possible, for it was feared that street pavements might cave in. When the first sections arrived at the Accelerator Building, a crowd had gathered in front of it. There were representatives of the City Departments that had issued permits to drive the heavy trucks on the streets. There were representatives of the utilities, gas and electricity, whose installations beneath the streets risked damage. There were policemen and photographers. Down the driveway to the Accelerator Building the trucks were held back with cables and winches: the enormous load made even the best brakes unsafe. It took three hours to bring a single section inside the building.

The magnet coils are made of a two-inch square copper bar with a hole in the middle to let water go through and cool the copper. The length of the bar is four and a half miles; the hole is big enough for a man's thumb to fit comfortably into it. The copper coils and the metal box were constructed at the New York Shipyard. They were too big to fit any train, so they were loaded on two barges, shipped by canal to Buffalo, transferred to a lake freighter, and brought to the lake front in Chicago. They reached the Accelerator Building on trucks, at the crack of dawn, under heavy police escort, over the widest streets, which had been cleared of parked cars.

Enrico took me to see the cyclotron when its shield was not yet complete and the magnet was not hidden from view. He made me leave my watch in an office upstairs, for the magnetic field would wreck it. Then we went down to the cyclotron pit.

The magnet did not look at all like the U-shaped magnet painted red that I use to pick up pins and needles when I sew. The cyclotron magnet was painted yellow. It was as tall as a small house, and its shape seemed intricate: I could not say which were its poles, despite Enrico's explanations.

Enrico made me climb on what seemed a wide ledge, the top of a wall twelve feet thick: it was almost a street! As I stood on this ledge, I watched the magnet perform some of its tricks. Enrico gave me a pocket knife saying:

"Hold it fast! Don't let it go!"

So far there had been no electric current in the coils. Now the current was switched on, and the magnet came alive. I felt the knife try hard to escape my hand. Then I noticed that Enrico's pocket was bulging out, that from its usual vertical position it was being drawn into a horizontal plane, as if an invisible ghost were trying to divest Enrico of his suit by pulling on his pocket. The magnet was attempting to steal Enrico's keys.

Together with the jolly spirit of a practical joker, a more malignant demon is housed in the magnet. Once there was a piece of concrete on the floor. It looked untidy, and Herbert Anderson, forgetting that the concrete was reinforced with steel, lifted it. The magnet grabbed it immediately, and Herbert's hand was crushed between stone and magnet.

The twelve-foot wall on which I stood was only part of the shield, a fact which duly impressed me. I returned to the Accelerator Building a few months later, when the cyclotron was entirely incased in huge beams of reinforced concrete. The ledge where I had stood and walked was no longer there. That wall was now higher and reached the ceiling of the shield.

While I was standing in front of another high wall, Enrico went to push a button. Silently part of the wall began to slide open. A block of concrete twelve feet thick, weighing sixty-nine tons, moved away. This block, called "John Marshall's door," provides an entrance to the cyclotron vault. John Marshall, who is justly proud of his door, explained its features to me. Safety devices, he said, make it impossible that a man be squashed in the doorway between the wall and the sixty-nine-ton block, or that someone be trapped inside the cyclotron vault; the cyclotron is automatically shut off and cannot emit radiations unless the door is closed. All possible precautions have been taken to protect workers.

As I stood spellbound watching the door silently close and open, I was suddenly twenty years back, in a room of the old physics building in Rome. The room was small, high ceilinged, and bare. Against one of its walls there was a crude-looking apparatus made of tall vertical bars supporting large metal balls. The apparatus reached the ceiling. I felt awfully disappointed when I saw it. Was this the high-voltage machine that Enrico and his friends had built and were proud of? The terms in which Enrico spoke of it were such that I had been eager to see it. So I had walked to the physics laboratory.

I did not feel like complimenting the apparatus; there was nothing attractive in it. But I had to say something. I saw an old table, probably a discarded desk, right in front of the machine. So I asked:

"What is this table doing here?"

The table, I learned, had a protective function. It prevented people from going too close to the apparatus and getting an electric shock. The physicists had put the table there after Amaldi had received a violent shock and was thrown down to the floor. Luckily for him, Segré was in the room and swiftly pulled the electric switch to the off position.

A little table . . . a twelve-foot thickness of concrete. . . .

The biggest of all gadgets that Fermi had ever seen, the cyclotron, was erected at the University of Chicago, right under his nose. Could it be possible that Enrico escaped its lure and left the entire construction to Herbert Anderson and John Marshall? Of course not!

Enrico built a small gadget for the big gadget.

While a cyclotron is working, nobody can approach it because it

Fermi's Streetcar

John Marshall and Herbert Anderson Look at the Cyclotron Shield

$$\frac{h}{mv} = 1{,}8 \times 10^{-8} \text{ cm}$$

A Lecture

Publifoto

Research

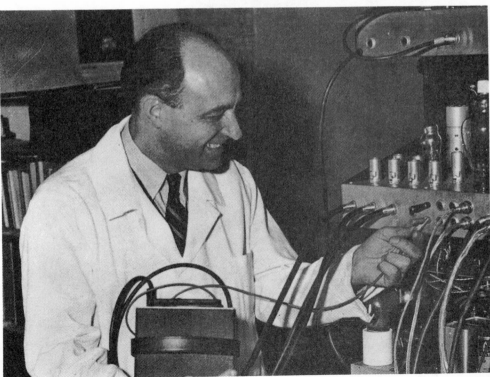

The Fermis and Son Giulio in Washington State, 1947

Capes

Time Goes By . . . 1953

emits dangerous radiation. Nobody can go near the vacuum box where protons are accelerated and experiments performed. If adjustments or changes in the position of experimental equipment are needed, if the target hit by the accelerated protons is to be moved, the cyclotron must be stopped. Unless, Enrico thought, the target could be placed on some sort of car and the car could run by itself.

So Enrico built what is now called "Fermi's trolley": a platform of lucite mounted on four wheels, looking as if they came from an erector set. But, Enrico said, offended, he built each single part with his own hands. The trolley burns no fuel and uses no electricity: it takes advantage of the cyclotron's magnetic field. The trolley does not need rails because its wheels fit the rim of the lower magnet pole. A target, or other small piece of equipment, fastened to the lucite platform, will ride the streetcar to any point along the rim of the magnetic pole, according to the whim of a push-button experimenter in the distant control room.

Fermi's trolley looks very neat, except for a sloppy joint of wires, conspicuous in front of the lucite platform. A man's nature never changes: Enrico has always given attention to performance and disregarded appearance.

What do physicists hope to achieve with their giant cyclotrons?

At the end of the war physicists found themselves in a strange situation. They had "harnessed nuclear energy," as the popular saying goes, but they knew little what the nucleus was like. Nuclei performed meekly at men's will, they split in two, they released the energy that was inside them. Nuclei were willing to do this very fast and to cause an atomic explosion, or slowly, in a controlled chain reaction. But they had not given up the secret of their structure.

This is what physicists claimed. But all scientists are greedy persons, who want to learn more and more, who are never satisfied with the state of their knowledge. The truth is that quite a few facts were known about nuclei: that they are made of protons and neutrons; that these are held together by very great forces; that these forces are different from any known so far. But the nature of nuclear forces escaped the physicists; it was a challenging riddle.

There was also another disturbing puzzle: if neutrons, protons, and electrons are the elementary constituents of all matter, no other particle should exist. But a number of other elementary particles were discovered in the cosmic radiation. Among them, the most talked about are the mesons. Where do the mesons fit into the picture of the nucleus?

Perhaps the most interesting feature of mesons is that, even before they were discovered, they had been postulated in an attempt to solve the mystery of the nuclear forces. In 1935 the Japanese physicist Hideki Yukawa proposed a theory to explain the salient characteristics of nuclear forces. To make his theory work, Yukawa had to assume the existence of particles that nobody had ever seen. They had to be of intermediate weight between electrons and nuclei, and for this reason they were called "mesons." Yukawa's theory would have been considered little more than an ingenious speculation if mesons, both positive and negative, had not been identified in the cosmic radiation shortly afterward. Yukawa was awarded the Nobel Prize for his meson theory in 1949.

The energy of the cosmic radiation is so spectacularly great that physicists did not hope to produce it with machines. But, they thought, they could perhaps build a cyclotron of sufficiently high energy to make the new elementary particles in the laboratory. High energy would also be of great help in trying to solve the riddle of nuclear forces. The best way to approach this problem is to shake protons loose and to see what happens at the moment they are torn apart. To do this, extremely high energies are needed. Hence the giant cyclotrons.

Lawrence's giant cyclotron in Berkeley was the first to be built and the first to make mesons in 1948. When the Chicago cyclotron was completed, it also made mesons.

Enrico had given thought to some theoretical aspects of mesons, but he had never experimented with them. Now he felt challenged by the new field of research that the cyclotron had opened for him.

Upon coming to Chicago from Los Alamos, Enrico had gone back to his old love: the neutrons. The new pile at the Argonne Laboratory provided a powerful source of these particles. Enrico often drove to Argonne and performed experiments both with slow and with fast neutrons. He studied their absorption and their optics; he improved past techniques.

"Had Enrico been like most physicists, he would have gone on working with neutrons, perfecting experiments in a field he knew well," Emilio Segré said recently. (Emilio came often to see us in Chicago and, like many of the old friends, he was happy when he "could pump a lot of physics out of Enrico.") "He would now be the king of the neutrons. What other physicist would want to learn new methods of research when he is fifty?"

Fermi's fiftieth birthday came on September 29, 1951, three months after the Chicago cyclotron started to work. So Fermi did learn new techniques at fifty; he shifted his allegiance from neutrons to mesons.

Enrico's interest in the cyclotron gave no sign of waning. So I was surprised when not long ago he went on a trip to Long Island to irradiate some photographic plates in the cosmotron. The cosmotron is a big machine, bigger than all giant cyclotrons, which was built at the Brookhaven National Laboratory in Long Island. The name "cosmotron" implies that the machine is in competition with the cosmos for the production of high energy.

"But it is far from it," Enrico said. "The cosmotron yields two billion electron volts, little more than four times the energy of our cyclotron. The cosmos imparts to the cosmic radiation energies millions of times greater than those obtained in the cosmotron."

Physicists are insatiable. A larger machine has recently started to work at Berkeley. It is called the "bevatron," and the first three letters of this word stand for "billion electron volts." It will produce up to six billion electron volts. CERN, the European center for nuclear research, has already placed orders for a synchroton, to be built at Geneva, capable of producing 25 billion electron volts. And I have heard talk of machines that physicists hope to build in the future, whose energies are expressed by figures too large for my memory to retain.

ACKNOWLEDGMENTS

All characters in this book are true. To them I owe an apology and the expression of my gratitude.

They may feel that I have not portrayed them as they would have liked, that I have stressed the amusing over the serious traits of their character. For this I apologize to them. Their willingness to talk about old times has helped me refresh my memory. For this I am deeply thankful.

A few persons have given me special assistance and I wish to offer special thanks to them:

To Dr. Cyril Smith, who gave me the idea for this book. "You should write your husband's biography," he told me. "I cannot," I answered. "My husband is the man I cook for and iron shirts for. How can I take him *that* seriously?" Yet the seed was sown, and it produced this book.

To Dr. Emilio Segrè for reading the parts of the manuscript dealing with scientific matters.

To Mrs. Earl Long and to Mrs. Morton Grodzins for their help and their suggestions.

Above all, to the members of my family, who have endured life with a writing housekeeper and have not complained.